Fusion

The Energy of the Universe

Second Edition

Fusion
The Energy of the Universe
Second Edition

Garry McCracken

and

Peter Stott

AMSTERDAM • BOSTON • HEIDELBERG • LONDON
NEW YORK • OXFORD • PARIS • SAN DIEGO
SAN FRANCISCO • SINGAPORE • SYDNEY • TOKYO
Academic Press is an imprint of Elsevier

Academic Press is an imprint of Elsevier
225 Wyman Street, Waltham, MA 02451, USA
The Boulevard, Langford Lane, Kidlington, Oxford, OX5 1GB, UK

Notices
Knowledge and best practice in this field are constantly changing. As new research and experience
broaden our understanding, changes in research methods, professional practices, or medical
treatment may become necessary.

Practitioners and researchers must always rely on their own experience and knowledge in
evaluating and using any information, methods, compounds, or experiments described herein. In
using such information or methods they should be mindful of their own safety and the safety of
others, including parties for whom they have a professional responsibility.

To the fullest extent of the law, neither the Publisher nor the authors, contributors, or editors,
assume any liability for any injury and/or damage to persons or property as a matter of products
liability, negligence or otherwise, or from any use or operation of any methods, products,
instructions, or ideas contained in the material herein.

Library of Congress Cataloging-in-Publication Data
Application submitted

British Library Cataloguing-in-Publication Data
A catalogue record for this book is available from the British Library

ISBN: 978-0-12-384656-3

For information on all Academic Press publications,
visit our website: http://store.elsevier.com

Printed and bound by CPI Group (UK) Ltd, Croydon, CR0 4YY

Transferred to digital print 2012

Working together to grow
libraries in developing countries

www.elsevier.com | www.bookaid.org | www.sabre.org

ELSEVIER   BOOK AID
            International   Sabre Foundation

To Pamela and Olga,
thank you for your continued encouragement
and patience.

Contents

1.   What Is Nuclear Fusion?

2.   Energy from Mass

3.   Fusion in the Sun and Stars

4.   Man-Made Fusion

Technical Summaries

These technical summaries, contained in shaded boxes, are supplements to the main text. They are intended for the more technically minded reader and may be bypassed by the general reader without loss of continuity.

Chapter 9

Chapter 10

Chapter 11

Chapter 12

Chapter 13

Foreword to the Second Edition

The imperative for fusion energy grows stronger every year. Concerns increase regarding global climate change, conflicts over natural resources, safety of massively deployed energy production, and the finite supply of energy resources. The need for a safe, carbon-free, abundant energy source is intense. Fusion is one of the very few options for such an energy source.

Over the past fifty years, scientists have developed a new field of science: plasma physics. The core of a fusion energy system is a plasma—a gas of charged particles—which is superhot, about ten times hotter than the core of the Sun. We now know how to produce, manipulate, and control this state of matter with remarkable finesse. The engineering and technology needed to control plasmas are also mature.

As a result of this fantastic accumulation of knowledge, fusion energy research is now at a turning point. With some reasonable clarity, we can see our way to the endpoint of commercial fusion energy. Both physics and engineering challenges surely remain. But the solutions to these challenges are sufficiently well-envisioned that roadmaps to fusion-on-the-grid have been developed in all nations' active fusion research. Across the world, scientists draw a similar conclusion that a full-scale demonstration power plant can be operating in some 20–25 years.

The current turning point is also marked by two new, landmark experiments—ITER and NIF (the National Ignition Facility). ITER is an experiment in magnetic fusion energy, in which plasmas are confined by strong magnetic fields. Remarkably, the governments of half the world's population have come together to design, construct, and operate the facility. ITER is designed, as an experiment, to generate about 500 million watts of fusion power for about 500 seconds. It will also produce, for the first time, a self-sustaining, burning plasma in which the plasma is self-heated by the power from the fusion reactions. ITER is under construction in Cadarache, France, and will begin operation around 2020. It will establish, to a large extent, the scientific and technological feasibility of a fusion power plant. NIF is an experiment in inertial fusion energy, recently brought into operation at the Lawrence Livermore National Laboratory in the US. NIF implodes a tiny pellet of frozen fusion fuel by immense lasers, causing a release of fusion energy in a fraction of a billionth of a second. Experiments are now underway to produce an ignited plasma in which fusion begins in the pellet core and propagates throughout the pellet.

This is an opportune moment to look back at the deep understanding of fusion plasma science that has accrued over the decades. This science has given us our designs of fusion power plants, but also has enormous impact on other fields of physics and technology, from the plasma cosmos to plasmas used to process computer chips. At this same moment, we can look ahead to how we will overcome the scientific hurdles to fusion. Thus, I am very pleased that the second, updated edition of this comprehensive and authoritative book by Garry McCracken and Peter Stott is now appearing, with new chapters on ITER and NIF.

The authors tell a fascinating scientific story in this book, in which both the history and the science of fusion unfold together, chronologically. Beginning with the early, mid-century days of fusion, we are taken up to the most recent developments. The treatment is scientifically comprehensive, covering topics from basic plasma physics to fusion materials science to engineering. And it treats both approaches to fusion—magnetic and inertial. A reader of this book will be rewarded by an understanding of why fusion is one of the most challenging scientific endeavors undertaken, how plasmas are rich with fascinating phenomena, how scientists across the world have met the challenges of fusion, and the road that lies ahead. Written by two veteran, accomplished fusion physicists, the book is an unusual opportunity to receive an insider's description of the scientific nuts and bolts, as well as the big picture, of this critical new energy source for the world.

Stewart Prager*

*Stewart Prager is currently Director of the Princeton Plasma Physics Laboratory and Professor of Astrophysical Sciences at Princeton University (2009 to present). Prior to that he was Professor of Physics at the University of Wisconsin – Madison (1977–2009). He was Chair of the DOE's Fusion Energy Sciences Advisory Committee (2006–2009), and Chair of the Division of Plasma Physics of the American Physical Society (APS) (1995–1996).

Fusion powers the stars and could in principle provide almost unlimited, environmentally benign, power on Earth. Harnessing fusion has proved to be a much greater scientific and technical challenge than originally hoped. In the early 1970s the great Russian physicist Lev Andreevich Artsimovich wrote that, nevertheless, "thermonuclear [fusion] energy will be ready when mankind needs it." It looks as if he was right and that that time is approaching. This excellent book is therefore very timely.

The theoretical attractions of fusion energy are clear. The raw fuels of a fusion power plant would be water and lithium. The lithium in one laptop computer battery, together with half a bath of water, would generate 200,000 kWh of electricity—as much as 40 tons of coal. Furthermore, a fusion power plant would not produce any atmospheric pollution (greenhouse gases, sulfur dioxide, etc.), thus meeting a requirement that is increasingly demanded by society.

The Joint European Torus (JET), at Culham in the United Kingdom, and the Tokamak Fusion Test Reactor (TFTR), at Princeton in the United States, have produced more than 10 MW (albeit for only a few seconds), showing that fusion can work in practice. The next step will be to construct a power-plant-size device called the International Thermonuclear Experimental Reactor (ITER), which will produce 500 MW for up to 10 minutes, thereby confirming that it is possible to build a full-size fusion power plant. The development of fusion energy is a response to a global need, and it is expected that ITER will be built by a global collaboration.

A major effort is needed to test the materials that will be needed to build fusion plants that are reliable and, hence, economic. If this work is done in parallel with ITER, a prototype fusion power plant could be putting electricity into the grid within 30 years. This is the exciting prospect with which this book concludes.

As early as 1920, it was suggested that fusion could be the source of energy in the stars, and the detailed mechanism was identified in 1938. It was clear by the 1940s that fusion energy could in principle be harnessed on Earth, but early optimism was soon recognized as being (in Artsimovich's words of 1962) "as unfounded as the sinner's hope of entering paradise without passing through purgatory." That purgatory involved identifying the right configuration of magnetic fields to hold a gas at over 100 million degrees Celsius (10 times hotter than the center of the Sun) away from the walls of its container. The solution of this challenging problem—which has been likened to holding a jelly with elastic bands—took a long time, but it has now been found.

Garry McCracken and Peter Stott have had distinguished careers in fusion research. Their book appears at a time when fusion's role as a potential ace of trumps in the energy pack is becoming increasingly recognized. I personally cannot imagine that sometime in the future, fusion energy will not be widely harnessed to the benefit of mankind. The question is when. This important book describes the exciting science of, the fascinating history of, and what is at stake in mankind's quest to harness the energy of the stars.

Chris Llewellyn Smith*

*Professor Sir Chris Llewellyn Smith FRS was Director UKAEA Culham Division (2003–2008), Head of the Euratom/UKAEA Fusion Association, and Chairman of the Consultative Committee for Euratom on Fusion (2003–2009). He was Director General of CERN (1994–1998) and Chairman of the ITER Council (2007–2009). He is currently Director of Energy Research at Oxford University, and President of the Council of SESAME (Synchrotron-light for Experimental Science and Applications in the Middle East).

Preface

Our aim in writing this book is to answer the frequently asked question "What is nuclear fusion?" In simple terms, *nuclear fusion* is the process in which two light atoms combine to form a heavier atom, in contrast to *nuclear fission*—in which a very heavy atom splits into two or more fragments. Both fusion and fission release energy and perhaps because of the similarity of the terms, they are sometimes confused. Nuclear fission is well known, but in fact nuclear fusion is much more widespread—fusion occurs continuously throughout the universe, and it is the process by which the Sun and the stars release energy and produce new elements from primordial hydrogen. It is a remarkable story.

There has been considerable research effort to use fusion to produce energy on Earth. Fusion would provide an environmentally clean and limitless source of energy. However, to release fusion energy, the fuel has to be heated to unbelievably high temperatures in the region of hundreds of millions of degrees Celsius—hotter in fact than the Sun. The obvious problem is how to contain such very hot fuel—clearly there are no material containers that will withstand such temperatures. There are two alternative ways to solve this problem. The first approach uses magnetic fields to form an insulating layer around the hot fuel. This approach, known as *magnetic confinement*, is now, after over 50 years of difficult research, at the stage where a prototype power plant could be built. The second approach is to compress and heat the fuel very quickly so that it burns and the fusion energy is released before the fuel has time to expand. This approach, known as *inertial confinement*, has made great progress over the last decade and is now close to achieving net energy output and demonstrating scientific feasibility.

In this book we present the complete story of fusion, starting with the development of the basic scientific ideas that led to the understanding of the role of fusion in the Sun and stars. We explain the processes of hydrogen burning in the Sun and the production of heavier elements in stars and supernovae. The development of fusion as a source of energy on Earth by both the magnetic- and inertial-confinement approaches is discussed in detail from the scientific beginnings to the construction of a fusion power plant. We briefly explain the principles of the hydrogen bomb and also review various false trails to fusion energy. The final chapter looks at fusion in the context of world energy needs.

The book has been structured to appeal to a wide readership. In particular, we hope it will appeal to readers with a general interest in science but little scientific background as well as to students who may find it useful as a supplement to more formal textbooks. The main text has been written with the minimum of scientific jargon and equations and emphasizes a simple and intuitive explanation of the scientific ideas. Additional material and more technical detail are included in the form of shaded "boxes" that will help the more serious student to understand some of the underlying physics and to progress to more advanced literature. However, the boxes are not essential reading, and we encourage the non-scientist to bypass them—the main text contains all that is needed to understand the story of fusion. We have tried to present the excitement of the scientific discoveries and to include brief portraits of some of the famous scientists who have been involved. The first edition has been translated and published in Czech, Japanese, Korean and Chinese.

In the second edition we have added considerable material to bring the story up to date. A chapter is included that is devoted to the design and building of the international fusion project ITER in southern France. ITER is planned to demonstrate the technical feasibility of magnetic fusion by achieving a power gain of 10 with an output of 500 MW. Another chapter has descriptions of the many large lasers that have been built and are being planned for further development of inertial-confinement systems. These include the National Ignition Facility (NIF) at Livermore, the Laser Mégajoule (LMJ) system near Bordeaux, and the FIREX facility at Osaka, Japan.

November 2011

Acknowledgments

In the course of writing this book we have drawn on the vast volume of published material relating to fusion in scientific journals and elsewhere, as well as on unpublished material and discussions with our colleagues. We have tried to give an accurate and balanced account of the development of fusion research that reflects the relative importance of the various lines that have been pursued and gives credit to the contributions from the many institutions in the countries that have engaged in fusion research. However, inevitably, there will be issues, topics, and contributions that some readers might feel deserved more detailed treatment. The contents of this book and the views expressed herein are the sole responsibility of the authors.

Special thanks are due to Kathryn Morrissey, Jeff Freeland, and their colleagues at Elsevier Academic Press Publishing for their patience and encouragement.

We would like to thank all of our colleagues who have helped and advised us in many ways. In particular, we are indebted to Steven Cowley, Mike Dunne, Roger Evans, Sabina Griffith, Guenter Janeschitz, Peter Norreys, Steven Rose, Mark Sherlock, Paul Thomas, and David Ward for their advice, constructive criticism, and valuable suggestions for improvement. We are particularly grateful to those colleagues who took a great deal of time and trouble to read drafts of some of the revised chapters. We are grateful also to our many colleagues who contributed to the first edition—Chris Carpenter, Jes Christiansen, Geoff Cordey, Richard Dendy, Jim Hastie, John Lawson, Ramon Leeper, Bruce Lipschultz, Kanetada Nagamine, Spencer Pitcher, Stephen Pitcher, Peter Stangeby, Neil Taylor, Fritz Wagner, John Wesson, Alan Wootton, and many others. Special thanks are due to Kathryn Morrissey and colleagues at Elsevier Academic Press Publishing for their patience and encouragement.

The authors thank the following organizations and individuals for granting permission to use the following figures: EFDA-JET (for Figures 1.2, 2.5, 3.8, 4.4, 4.5, 4.6, 4.8, 5.2, 5.4, 9.2, 9.3, 9.8, 10.1, 10.2, 10.3, 10.5, 10.6, 10.7, 10.8, 10.11, and 13.1); UK Atomic Energy Authority (for Figures 5.3, 5.5, 5.7, 5.8, 13.3, 13.4, 13.5, and 14.6; also for permission to use the archive photographs in Figures 5.1, 5.9, 5.10, and 9.1); Lawrence Livermore National Laboratory (for Figures 7.3b, 7.6, 7.7, 7.8, 7.9, 12.1, 12.2, 12.3, 12.4, 12.5, and 12.7); Sandia National Laboratories (for Figures 7.11 and 7.12); Max-Planck-Institut für Plasmaphysik (for Figures 9.6, 10.12, and 10.13); Princeton Plasma Physics Laboratory (for Figure 10.4); ITER (for Figures 11.1 and 11.3 through 11.11); John Lawson (for Figure 4.7); Institute for Laser Technology, Osaka University (for Figures 12.8 and 12.9); Japan Atomic Energy Agency, Naka Fusion Institute (for Figure 10.10); CEA (for Figure 12.6); The Royal Society (for Figure 2.3); Master and Fellows of Trinity College, Cambridge (for Figure 2.4—used by permission); NASA image from Solarviews.com (for Figure 3.3); AIP Emilio Segrè Visual Archives (for Figures 3.4 and 5.6); Figure 3.6 © Malin/IAC/RGO and Photograph by David Malin; Figure 3.7 © Australian Astronomical Observatory and Photograph by David Malin from AAT plates; Lawrence Berkeley National Laboratory, courtesy AIP Emilio Segre Visual Archives (for Figure 6.1); VNIIEF Museum and Archive, Courtesy AIP Emilion Segrè Visual Archives, *Physics Today* Collection (for Figure 6.4); The Lebedev Physical Institute of the Russian Academy of Sciences (for Figure 7.5); Figure 3.1 originally appeared in *The Life of William Thomson*, Macmillan, 1910. The copyright for these materials remains with the original copyright holders. Every effort has been made to contact all copyright holders. The authors would be pleased to make good in any future editions any errors or omissions that are brought to their attention.

Some figures have been redrawn or modified by the authors from the originals or are based on published data. Figure 3.5 is adapted from *Nucleosynthesis and Chemical Evolution of Galaxies* by B.E.J. Pagel (Cambridge University Press, Cambridge, UK, 1997); Figure 6.2 is adapted from *Dark Sun: The Making of the Hydrogen Bomb* by R. Rhodes (Simon and Schuster, New York, NY, 1996); Figures 7.1, 7.2, and 7.10 are adapted from J. Lindl, *Physics of Plasmas*, **2** (1995) 3939; Figure 8.1 is adapted from *Too Hot to Handle: The Story of the Race for Cold Fusion* by F. Close (W. H. Allen Publishing, London, 1990); Figure 8.2 is adapted from K. Ishida *et al.*, *J. Phys.* **G 29** (2003) 2043; Figure 9.4 is adapted from an original drawing by General Atomics; Figure 9.5 is adapted from *L'Energie des Etoiles, La Fusion Nucleaire Controlée* by P-H. Rebut (Editions Odile Jacob, Paris, 1999); Figure 11.2 is based in part on data published by R. J. Hawryluk *et al.* in *Nucl. Fusion* **49** (2009) 065012; Figures 13.6 and 13.7 are based in part on data published by M. Dunne *et al.* (LLML-CONF-463771); Figures 14.1, 14.3, and 14.4 are based in part on data published by the International Energy Agency in *Key World Energy Statistics 2010*; Figure 14.2 is adapted from the World Energy Council Report, *Energy for Tomorrow's World: The Realities, the Real Options and the Agenda for Achievement* (St. Martin's Press, New York, NY, 1993); Figure 14.5 has been modified from a figure that was created by Robert A. Rohde from public data and that is incorporated into the Global Warming Art project; and Figure 14.7 is based in part on data published by the US Department of Energy. We are particularly grateful to Stuart Morris and his staff, who drew or adapted many of the figures specifically for this book.

Chapter 1

What Is Nuclear Fusion?

1.1 The Alchemists' Dream

In the Middle Ages, the alchemists' dream was to turn lead into gold. The only means of tackling this problem were essentially chemical ones, and these were doomed to failure. During the 19th century, the science of chemistry made enormous advances, and it became clear that lead and gold are different elements that cannot be changed into each other by chemical processes. However, the discovery of radioactivity at the very end of the 19th century led to the realization that sometimes elements do change spontaneously (or transmute) into other elements. Later, scientists discovered how to use high-energy particles, either from radioactive sources or accelerated in the powerful new tools of physics that were developed in the 20th century, to induce artificial *nuclear transmutation* in a wide range of elements. In particular, it became possible to split atoms (the process known as *nuclear fission*) or to combine them (the process known as *nuclear fusion*). The alchemists (Figure 1.1) did not understand that their quest was impossible with the tools they had at their disposal, but in one sense it could be said that they were the first people to search for nuclear transmutation.

FIGURE 1.1 An alchemist in search of the secret that would change lead into gold. Because alchemists had only chemical processes available, they had no hope of making the nuclear transformation required. *From a painting by David Teniers the younger, 1610–1690.*

Fusion, Second Edition.

What the alchemists did not realize was that nuclear transmutation was occurring before their very eyes, in the Sun and in all the stars of their night sky. The processes in the Sun and stars, especially the energy source that had sustained their enormous output for eons, had long baffled scientists. Only in the early 20th century was it realized that nuclear fusion is the energy source that runs the universe and that simultaneously it is the mechanism responsible for creating all the different chemical elements around us.

1.2　The Sun's Energy

The realization that the energy radiated by the Sun and stars is due to nuclear fusion followed three main steps in the development of science. The first was Albert Einstein's famous deduction in 1905 that mass can be converted into energy. The second step came a little over 10 years later, with Francis Aston's precision measurements of atomic masses, which showed that the total mass of four hydrogen atoms is slightly larger than the mass of one helium atom. These two key results led Arthur Eddington and others, around 1920, to propose that mass could be turned into energy in the Sun and the stars if four hydrogen atoms combine to form a single helium atom. The only serious problem with this model was that, according to classical physics, the Sun was not hot enough for nuclear fusion to take place. It was only after *quantum mechanics* was developed in the late 1920s that a complete understanding of the physics of nuclear fusion became possible.

Having answered the question as to where the energy of the universe comes from, physicists started to ask how the different atoms arose. Again fusion was the answer. The fusion of hydrogen to form helium is just the start of a long and complex chain. It was later shown that three helium atoms can combine to form a carbon atom and that all the heavier elements are formed in a series of more and more complicated reactions. Nuclear physicists played a key role in reaching these conclusions. By studying the different nuclear reactions in laboratory accelerators, they were able to deduce the most probable reactions under different conditions. By relating these data to the astrophysicists' models of the stars, a consistent picture of the life cycles of the stars was built up and the processes that give rise to all the different atoms in the universe were discovered.

1.3　Can We Use Fusion Energy?

When fusion was identified as the energy source of the Sun and the stars, it was natural to ask whether the process of turning mass into energy could be demonstrated on Earth and, if so, whether it could be put to use for man's benefit. Ernest Rutherford, the famous physicist and discoverer of the structure of the atom, made this infamous statement to the British Association for the Advancement of Science in 1933: "We cannot control atomic energy to an extent that would be of any use commercially, and I believe we are not ever likely to do so." It was one of the few times when his judgment proved wanting. Not everybody shared Rutherford's view; H. G. Wells had predicted the use of nuclear energy in a novel published in 1914.[1]

The possibility of turning nuclear mass into energy became very much more real in 1939, when Otto Hahn and Fritz Strassman demonstrated that the uranium atom could be split by bombarding uranium with neutrons, with the release of a large amount of energy. This was fission. The story of the development of the fission chain reaction, fission reactors, and the atom bomb has been recounted many times. The development of the hydrogen bomb and the quest for fusion energy proved to be more difficult. There is a good reason for this. The uranium atom splits when bombarded with *neutrons*. Neutrons, so called because they have no electric charge, can easily penetrate the core of a uranium atom, causing it to become unstable and to split. For fusion to occur, two hydrogen atoms have to get so close to each other that their cores can merge; but these cores carry strong electric charges that hold them apart. The atoms have to be hurled together with sufficiently high energy to make them fuse.

1.4 Man-Made Suns

The fusion reaction was well understood by scientists making the first atomic (fission) bomb in the Manhattan Project. However, although the possibility that fusion could be developed as a source of energy was undoubtedly discussed, no practical plans were put forward. Despite the obvious technical difficulties, the idea of exploiting fusion energy in a controlled manner was seriously considered shortly after World War II, and research was started in the UK at Liverpool, Oxford, and London universities. One of the principal proponents was George Thomson, the Nobel Prize-winning physicist and son of J. J. Thomson, the discoverer of the electron. The general approach was to try to heat hydrogen gas to a high temperature so that the colliding atoms have sufficient energy to fuse together. By using a magnetic field to confine the hot fuel, it was thought that it should be possible to allow adequate time for the fusion reactions to occur. Fusion research was taken up in the UK, the US, and the Soviet Union under secret programs in the 1950s and subsequently, after being declassified in 1958, in many of the technically advanced countries of the world. The most promising reaction is that between the two rare forms of hydrogen, called *deuterium* and *tritium*. Deuterium is present naturally in water and is therefore readily available. Tritium is not available naturally and has to be produced *in situ* in the power plant. This can be done by using the products of the fusion reaction to interact with the light metal lithium

---

1. *Atomic* energy and *nuclear* energy are the same thing.

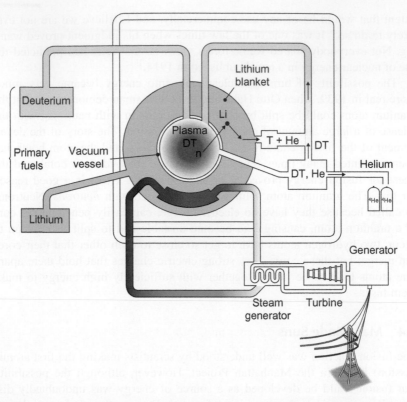

FIGURE 1.2 Schematic diagram of a proposed nuclear fusion power plant. The deuterium and tritium fuel burns at a very high temperature in the central reaction chamber. The energy is released as charged particles, neutrons, X-rays, and ultraviolet radiation and it is absorbed in a lithium blanket surrounding the reaction chamber. The neutrons convert the lithium into tritium fuel. A conventional steam-generating plant is used to convert the nuclear energy to electricity. The waste product from the nuclear reaction is helium.

in a layer surrounding the reaction chamber in a *breeding cycle*. Thus the basic fuels for nuclear fusion are lithium and water, both readily and widely available. Most of the energy is released as heat that can be extracted and used to make steam and drive turbines, as in any conventional power plant. A schematic diagram of the proposed arrangement is shown in Figure 1.2. The problem of heating and containing the hot fuel with magnetic fields (*magnetic-confinement fusion*) turned out to be much more difficult than at first envisaged.

However, research on the peaceful use of fusion energy was overtaken in a dramatic way with the explosion of the hydrogen bomb in 1952. This stimulated a second approach to controlled fusion, based on the concept of heating the fuel to a sufficiently high temperature very quickly before it has time to escape. The invention of the laser in 1960 provided a possible way to do this; lasers can focus intense bursts of energy onto small targets. The idea is to rapidly heat and compress small fuel pellets or capsules in a series of mini-explosions. This is called *inertial confinement* because the fusion fuel is confined only by its own inertia. Initially, the expertise was limited to those countries that already had nuclear weapons, and some details still remain secret, although other countries have now taken it up for purely peaceful purposes. Apart from the heating and confinement of the fuel, the method of converting fusion energy into electricity will be very similar to that envisaged for magnetic confinement.

1.5   The Rest of the Story

The considerable scientific and technical difficulties encountered by the magnetic and inertial-confinement approaches have caused these programs to stretch over many years. The quest for fusion has proved to be one of the most difficult challenges faced by scientists. After many years, the scientific feasibility of thermonuclear fusion via the magnetic-confinement route has been demonstrated, and inertial-confinement experiments are expected to reach a similar position soon. Developing the technology and translating these scientific achievements into power plants that are economically viable will be a major step that will require much additional time and effort. Some have hoped that they could find easy ways to the rewards offered by fusion energy. This line of thinking has led to many blind alleys and even to several false claims of success, the most widely publicized being the so-called "cold fusion" discoveries that are described in Chapter 8.

Chapter 2

Energy from Mass

2.1 Einstein's Theory

Energy is something with which everyone is familiar. It appears in many different forms, including electricity, light, heat, chemical energy, and motional (or *kinetic*) energy. An important scientific discovery in the 19th century was that *energy is conserved*. This means that energy can be converted from one form to another, but the total amount of energy must stay the same. *Mass* is also very familiar, though sometimes it is referred to, rather inaccurately, as weight. On the Earth's surface, mass and weight are often thought of as being the same thing, and they do use the same units—something that weighs 1 kilogram has a mass of 1 kilogram—but strictly speaking *weight* is the force that a mass experiences in the Earth's gravity. An object always has the same mass, even though in outer space it might appear to be weightless. Mass, like energy, is conserved.

The extraordinary idea that mass and energy are equivalent was proposed by Albert Einstein (Figure 2.1) in a brief three-page paper published in 1905. At that time, Einstein was a young man who was virtually unknown in the scientific world. His paper on the equivalence of mass and energy was followed soon after by three seminal papers—on the photoelectric effect, on Brownian motion, and on special relativity—all published in the same year. Henri Becquerel had discovered radioactivity 10 years previously. Using simple equations and application of the laws of conservation of energy and momentum, Einstein argued that the atom left after a radioactive decay event had emitted energy in the form of radiation must be less massive than the original atom. From this analysis he deduced that "If a body gives off the energy E in the form of radiation, its mass diminishes by E/c^2." He went on to say, "It is not impossible that with bodies whose energy content is variable to a high degree (e.g., radium salts) the theory may be successfully put to the test."

Einstein's deduction is more commonly written as $E = mc^2$, probably the most famous equation in physics. It states that mass is another form of energy and that energy equals mass multiplied by the velocity of light squared. Although it took a long time to get experimental proof of this entirely theoretical prediction, we now know that it was one of the most significant advances ever made in science.

Fusion, Second Edition.

FIGURE 2.1   Photograph of Albert Einstein in 1921—the year in which he received the Nobel Prize in Physics. Einstein (1879–1955) had graduated in 1901 and had made a number of applications for academic jobs, without success. He eventually got a job as technical expert, third class, in the Swiss patent office in Berne, which meant that he had to do all his research in his spare time.

2.2   Building Blocks

To see how Einstein's theory led to the concept of fusion energy, we need to go back to the middle of the 19th century. As the science of chemistry developed, it became clear that everything is built up from a relatively small number of basic components, called *elements*. At that time, about 50 elements had been identified, but we now know that there are around 100. As information accumulated about the different elements, it became apparent that there were groups of them with similar properties. However, it was not clear how they were related to each other until the *Periodic Table* was proposed by the Russian chemist Dmitri Mendeleev. In 1869 he published a table in which the elements were arranged in rows, with the lightest elements, such as hydrogen, in the top row and the heaviest in the bottom row. Elements with similar physical and chemical properties were placed in the same vertical columns. The table was initially imperfect, mainly because of inaccuracies in the data and because some elements had not yet been discovered. In fact, gaps in Mendeleev's table stimulated the search for, and the discovery of, new elements.

Each element consists of tiny units called *atoms*. Ernest Rutherford deduced in 1911 that atoms have a heavy core called the *nucleus* that has a positive electric charge. A cloud of lighter particles called *electrons* with a negative electric charge surrounds the nucleus. The negative electric charges of the electrons and the positive charge of the nucleus balance each other so that the atom overall has no net electric charge. The number of positive charges and electrons is different for each element, and this determines the element's chemical properties and its position in Mendeleev's table. Hydrogen is the

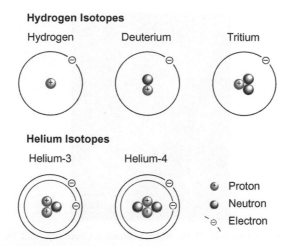

FIGURE 2.2 Structure of the different atoms of hydrogen and helium. Atoms with the same number of protons and different numbers of neutrons are known as isotopes of the same element.

simplest element, with just one electron in each atom; helium is next, with two electrons; lithium has three; and so on down to uranium, which, with 92 electrons, is the heaviest naturally occurring element. Schematic diagrams of the structure of the atoms of hydrogen and helium are shown in Figure 2.2.

The chemists developed skilled techniques to measure the average mass of the atoms of each element—the *atomic mass* (which is also known as the *atomic weight*). Many elements were found to have atomic masses that were close to being simple multiples of the atomic mass of hydrogen, and this suggested that, in some way that was not understood at that time, hydrogen might be a sort of building block for the heavier elements. To take some common examples, the atomic mass of carbon is approximately 12 times that of hydrogen, and the atomic mass of oxygen is 16 times that of hydrogen. There were some puzzling cases, however, that did not fit the general pattern. For example, repeated measurements of the atomic mass of chlorine gave a value of 35.5 times that of hydrogen.

The next significant step in the story was the direct measurement of the masses of individual atoms. During the period 1918–1920 at Cambridge University, UK, Francis Aston (Figure 2.3) built an instrument (Box 2.1) that could do this. Having studied chemistry at Birmingham, Aston had become interested in passing currents through gases in low-pressure discharge tubes. In 1910 he was invited to the Cavendish laboratory at Cambridge by J. J. Thomson, who was studying positive rays, also by using discharge tubes. Aston helped Thomson to set up an apparatus for measuring the mass-to-charge ratio of the positive species in the discharge. After World War I, Aston returned to Cambridge and started to measure the mass of atoms by a new method that was a great improvement on the Thomson apparatus. He subjected

FIGURE 2.3   Francis Aston (1877–1945), Nobel Laureate in Chemistry 1922. He started his scientific career by setting up a laboratory in a barn at his parents' home while still a schoolboy.

the atoms to an electric discharge, which removed one or more of their electrons. This left the nucleus surrounded with a depleted number of electrons and thus with a net positive electric charge—this is known as an *ion*. The ions were accelerated by an electric field to a known energy and then passed through a magnetic field. By measuring the amount by which they were deflected in the magnetic field, Aston was able to determine the mass of the atoms. The instrument was dubbed a *mass spectrograph* because the beams of ions were dispersed into a spectrum in a similar way to that in which a prism disperses light. Aston was a brilliant experimentalist with an obsession for accuracy. Gradually he developed greater and greater precision until he was able to determine the mass of an atom to an accuracy of better than one part in a thousand. These precision measurements yielded a number of entirely unexpected results. It is a good example of pure scientific curiosity leading eventually to valuable practical information.

Aston found that some atoms that are chemically identical could have different masses. This resolved the puzzle about the atomic weight of chlorine. There are two types of chlorine atom; one type is about 35 times heavier than hydrogen, and the other is about 37 times heavier. The relative abundance of the two types (75% have mass 35 and 25% have mass 37) gives an average of 35.5—in agreement with the chemically measured atomic mass. Likewise, Aston found that the mass of a small percentage (about 0.016%) of hydrogen atoms is almost double that of the majority. Atoms with different masses but the same chemical properties are called *isotopes*.

The reason for the difference in mass between isotopes of the same element was not understood until 1932, when James Chadwick discovered the neutron. It was then realized that the nucleus contains two types of atomic particle: *protons*, with a single unit of positive electric charge, and *neutrons*, with

Box 2.1 The Mass Spectrograph

The Aston mass spectrograph was an important development in the study of atomic masses. Starting by ionizing atoms either in an electric discharge or by an electron beam, a beam of ions is produced that is accelerated in an electric field to a fixed energy, eV, determined by the equation

$$\tfrac{1}{2}mv^2 = eV$$

where m, v, and e are the mass, velocity, and charge of the ions and V is the voltage through which the ions are accelerated.

The ions then pass into a uniform magnetic field, which exerts a force on them at right angles to the direction of the field and to the direction of the ion. The magnetic field (B) provides the centripetal force on the ions, forcing them to follow a circular path whose radius (r) is given by the equation

$$mv^2/r = Bev$$

Because all the ions have the same energy, the radius r of their circular path depends on their mass-to-charge ratio. The ions are thus dispersed spatially, rather as light is dispersed by a prism. One of the principal advantages of the geometry chosen by Aston is that the ions with the same ratio of mass to charge are spatially focused at the detector, thus optimizing the efficiency with which the ions are collected.

Many variations of the mass spectrograph (a variation using electrical detection is known as the mass spectrometer) have been developed and are widely used for routine analysis of all types of samples. One interesting application is its use for archaeological dating by measuring the ratio of the abundances of two isotopes of an element. If one isotope is radioactive, the age of a sample can be deduced. This analysis is often applied to ^{14}C and to the rubidium isotopes ^{85}Rb and ^{87}Rb, but other elements can be used, depending on the age of the sample being analyzed.

no electric charge. The number of protons equals the number of electrons, so an atom is overall electrically neutral. All isotopes of the same element have the same number of protons and the same number of electrons, so their chemical properties are identical. The number of neutrons can vary. For example, chlorine always has 17 protons, but one isotope has 18 neutrons and the other has 20. Likewise, the nucleus of the most common isotope of hydrogen consists of a single proton; the heavier forms, *deuterium* and *tritium*, have one proton with one and two neutrons, respectively, as shown in Figure 2.2. Protons and neutrons have very similar masses (the mass of a neutron is 1.00138 times the mass of a proton), but electrons are much lighter (a proton is about 2000 times the mass of an electron). The total number of protons and neutrons therefore determines the overall mass of the atom.

2.3   Something Missing

The most surprising result from Aston's work was that the masses of individual isotopes are not exactly multiples of the mass of the most common isotope of hydrogen; they are consistently very slightly lighter than expected. Aston had defined his own scale of atomic mass by assigning a value of precisely 4 to helium. On this scale, the mass of the light isotope of hydrogen is 1.008, so the mass of a helium atom is only 3.97 times, rather than exactly 4 times, the mass of a hydrogen atom. The difference is small, but Aston's reputation for accuracy was such that the scientific world was quickly convinced by his results.

The significance of this result was quickly recognized by a number of people. One was Arthur Eddington (Figure 2.4), now considered to be the most distinguished astrophysicist of his generation. He made the following remarkably prescient statement at the British Association for Advancement of Science meeting in Cardiff in 1920, only a few months after Aston had published his results.

Aston has further shown conclusively that the mass of the helium atom is less than the sum of the masses of the four hydrogen atoms which enter into it and in this at least the chemists agree with him. There is a loss of mass in the synthesis amounting to 1 part in 120, the atomic weight of hydrogen being 1.008 and that of helium just 4.00.... . Now mass cannot be annihilated and the deficit can only represent the mass of the electrical energy liberated when helium is made out of hydrogen. If 5% of a star's mass consists initially of hydrogen atoms, which are gradually being combined to form more complex elements, the total heat liberated will more than suffice for our demands, and we need look no further for the source of a star's energy.

If, indeed, the subatomic energy in the stars is being freely used to maintain their furnaces, it seems to bring a little nearer to fulfillment our dream of controlling this latent power for the well-being of the human race—or for its suicide.

FIGURE 2.4   Arthur Eddington (1882–1944) from the drawing by Augustus John.

Eddington had realized that there would be a mass loss if four hydrogen atoms combined to form a single helium atom. Einstein's equivalence of mass and energy led directly to the suggestion that this could be the long-sought process that produces the energy in the stars! It was an inspired guess, all the more remarkable because the structure of the nucleus and the mechanisms of these reactions were not fully understood. Moreover, it was thought at that time that there was very little hydrogen in the Sun, which accounts for Eddington's assumption that only 5% of a star's mass might be hydrogen. It was later shown that, in fact, stars are composed almost entirely of hydrogen.

In fact, according to the classical laws of physics, the processes envisaged by Eddington would require much higher temperatures than exist in the Sun. Fortunately, a new development in physics known as *quantum mechanics* soon provided the answer and showed that fusion can take place at the temperatures estimated to occur in the Sun. The whole sequence of processes that allows stars to emit energy over billions of years was explained in detail by George Gamow, by Robert Atkinson and Fritz Houtermans in 1928, and by Hans Bethe in 1938.

The question as to who first had the idea that fusion of hydrogen into helium was the source of the Sun's energy led to some bitter disputes, particularly between Eddington and James Jeans. Each thought they had priority, and they were on bad terms for many years as a result of the dispute.

As the techniques of mass spectroscopy were refined and made increasingly more accurate, detailed measurements were made on every isotope of every element. It was realized that many isotopes are lighter than would be expected by simply adding up the masses of the component parts of their nuclei—the protons and neutrons. Looked at in a slightly different way, each proton or neutron when combined into a nucleus has slightly less mass than when it exists as a free particle. The difference in mass per nuclear particle is called the *mass defect*, and, when multiplied by the velocity of light squared, it represents the amount of energy associated with the forces that hold the nucleus together.

These data are usually plotted in the form of a graph of the energy equivalent of the mass defect plotted against the total number of protons and neutrons in the nucleus (the atomic mass). A modern version of this plot is shown in Figure 2.5. While there are some irregularities in the curve at the left-hand side, for the lightest isotopes, most of the curve is remarkably smooth. The most important feature is the minimum around mass number 56. Atoms in this range are the most stable. Atoms to either side have excess mass that can be released in the form of energy by moving toward the middle of the curve, that is, if two lighter atoms join to form a heavier one (this is *fusion*) or a very heavy atom splits to form lighter fragments (this is *fission*).

It turns out that splitting the heavy atoms is very much the easier task, but the discovery of how it can be done was quite accidental. After the neutron had been discovered, it occurred to a number of groups to bombard uranium,

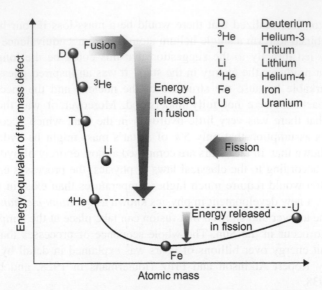

FIGURE 2.5   The energy equivalent of the mass defect of the elements plotted as a function of their atomic mass. Iron (Fe) with atomic mass 56 has the smallest mass defect. The amount of energy released when one atom is transmuted to another, either by fusing light atoms or by splitting heavy ones, is equal to the difference in their masses.

the heaviest naturally occurring element with an atomic mass of about 238, with neutrons in order to try to make even heavier *transuranic elements*. The amount of any new element was expected to be exceedingly small, and very sensitive detection techniques were required. Some genuine transuranic elements were detected, but there were some reaction products that did not fit the expectations. In 1939, Otto Hahn and Fritz Strassman performed a series of experiments that showed conclusively that these unexplained products were actually isotopes of barium and lanthanum that have mass numbers of 139 and 140, respectively, roughly half the mass of the uranium target nuclei. The only possible explanation was that the neutron bombardment of the uranium had induced fission in the uranium nucleus, causing it to split into two approximately equal parts. Moreover, it turned out that additional neutrons were released in the process. It was quickly realized that these neutrons could in principle induce further fission reactions, leading to a chain reaction. This led to the building of the first atomic reactor by Enrico Fermi in Chicago in 1943 and the development of the atomic bomb in Los Alamos.

Chapter 3

Fusion in the Sun and Stars

3.1 The Source of the Sun's Energy

At the beginning of the 20th century there was no convincing explanation for the enormous amount of energy radiated by the Sun. Although physics had made major advances during the previous century and many people thought that there was little of the physical sciences left to be discovered, they could not explain how the Sun could continue to release energy, apparently indefinitely. The law of energy conservation requires that there be an internal energy source equal to that radiated from the Sun's surface. The only substantial sources of energy known at that time were wood and coal. Knowing the mass of the Sun and the rate at which it radiated energy, it was easy to show that if the Sun had started off as a solid lump of coal it would have burnt out in less than 2000 years. It was clear that this was much too short—the Sun had to be older than the Earth, and the Earth was known to be older than 2000 years—but just how old was the Earth?

Early in the 19th century, most geologists had believed that the Earth might be indefinitely old. This idea was disputed by the distinguished physicist William Thomson, who later became Lord Kelvin (Figure 3.1). His interest in this topic began in 1844 while he was still a Cambridge undergraduate. It was a topic to which he returned repeatedly and that drew him into conflict with other scientists, such as John Tyndall, Thomas Huxley, and Charles Darwin. To evaluate the age of the Earth, Kelvin tried to calculate how long it had taken the planet to cool from an initial molten state to its current temperature. In 1862 he estimated the Earth to be 100 million years old. To the chagrin of the biologists, Kelvin's calculations for the age of the Earth did not allow enough time for evolution to occur. Over the next four decades, geologists, paleontologists, evolutionary biologists, and physicists joined in a protracted debate about the age of the Earth. During this time, Kelvin revised his figure down to between 20 million and 40 million years. The geologists tried to make quantitative estimates based on the time required for the deposition of rock formations or the time required to erode them, and they concluded that the Earth must be much older than Kelvin's values. However, too many unknown factors were required for such calculations, and they were generally considered unreliable. In the first edition of his book, *The Origin of Species*, Charles Darwin calculated the age of the Earth to be 300 million years, based on the time estimated to erode the Weald, a valley between the North and

Fusion, Second Edition.

FIGURE 3.1   William Thomson, later Lord Kelvin (1824–1907). Kelvin was one of the pioneers of modern physics, developing thermodynamics. He had a great interest in practical matters and helped to lay the first transatlantic telegraph cable.

South Downs in southern England. This was subjected to so much criticism that Darwin withdrew this argument from subsequent editions.

The discrepancy between the estimates was not resolved until the beginning of the 20th century, when Ernest Rutherford realized that radioactivity (discovered by Henri Becquerel in 1896, well after Kelvin had made his calculations) provides the Earth with an internal source of heat that slows down the cooling. This process makes the Earth older than was originally envisaged; current estimates suggest that our planet is at least 4.6 billion years old. Radioactivity, as well as providing the additional source of heat, provides an accurate way of measuring the age of the Earth by comparing the amounts of radioactive minerals in the rocks. The age of the Earth put a lower limit on the age of the Sun and renewed the debate about the source of the Sun's energy— What mechanism could sustain the Sun's output for such a long period of time? It was not until the 1920s, when Eddington made his deduction that fusion of hydrogen was the most likely energy source, and later, when quantum theory was developed, that a consistent explanation became possible.

3.2   The Solar Furnace

Hydrogen and helium are by far the most common elements in the universe, and together they account for about 98% of all known matter. There is no significant amount of hydrogen or helium in the gaseous state on Earth (or on Mars or Venus) because the gravity of small planets is too weak to keep these light atoms attached; they simply escape into outer space. All of the Earth's hydrogen is combined with oxygen as water, with carbon as hydrocarbons, or with other elements in the rocks. However, the Sun, whose gravity is much

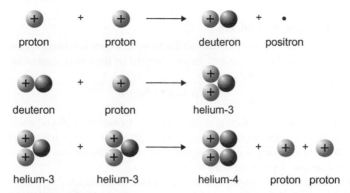

FIGURE 3.2   The three reactions that convert hydrogen to helium in the Sun. In the first stage two protons combine together. One proton converts to a neutron and the overall charge balance is conserved through emission of a *positron*, a particle with the same mass and properties as an electron but with positive electric charge.

stronger, consists almost entirely of hydrogen. The presence of hydrogen in the Sun and the stars can be measured directly from spectroscopic observations, since every atom emits light with characteristic *wavelengths* (or colors) that uniquely identify it.

Although there is plenty of hydrogen in the Sun for nuclear fusion, how can we know that conditions are right for fusion to occur? The temperature and density of the Sun can be determined by a combination of experimental observations using spectroscopy and by theoretical calculations. The most likely fusion reactions can be deduced from studies of nuclear reactions in the laboratory, using particle accelerators. The energy release in the Sun involves the conversion of four protons into a helium nucleus. However, this does not happen in a single step. First, two protons combine together and one of them converts into a neutron to form a nucleus of the heavy isotope of hydrogen known as *deuterium*. The deuterium nucleus then combines with another proton to form the light helium isotope known as *helium-3*. Finally, two helium-3 nuclei combine to form *helium-4*, releasing two protons in the process. Overall, four protons are converted into one helium nucleus. Energy is released because the helium nucleus has slightly less mass than the original four protons from which it was formed, as discussed in Chapter 2. The structure of the different nuclei is illustrated in Figure 2.2. The reactions are shown schematically in Figure 3.2, with the original nuclei on the left-hand side and the products on the right. Energy is released in each stage of the reactions. The total amount of energy released for each conversion of four hydrogen nuclei into a helium nucleus is about 10 million times more than is produced by the chemical reaction when hydrogen combines with oxygen and burns to form water. This enormous difference between the energy released by nuclear reactions compared to chemical reactions explains why fusion can sustain the Sun

Box 3.1 The Neutrino Problem

The first stage of the chain of reactions in the Sun—that between two protons—also releases a particle called a *neutrino* (usually denoted by the Greek symbol ν):

$$p + p \rightarrow D + {}^+e + \nu$$

Neutrinos were first predicted by Enrico Fermi, to explain a discrepancy in the energy measured in the process known as beta decay. Neutrinos have a very low probability of undergoing a nuclear reaction with other matter and therefore have a high probability of escaping from the Sun.

It was realized that if these neutrinos could be detected on Earth, they could be useful in determining the rate of fusion reactions in the Sun and also would help to answer other questions of fundamental physics. By devising very sensitive detectors and placing them deep underground, where they are shielded from other forms of radiation, it proved possible to detect solar neutrinos. It came as a surprise that the neutrino flux detected was only about one-third of that expected from other estimates of the rate of fusion reactions in the Sun. It was predicted that there should be three types of neutrino: the *electron neutrino*, the *muon neutrino*, and the *tau neutrino*. The product of the proton-proton reaction in the Sun is an electron neutrino, but it gradually began to be suspected that the electron neutrinos might be changing into one of the other types on the way from the Sun to the Earth. The puzzle was solved by measurements at a laboratory built some 2000 m underground in a nickel mine in Canada. The Sudbury Neutrino Observatory (SNO) started operating in 1999 with a detector using 1000 tons of heavy water. These detectors are able to measure the total neutrino flux as well as the electron neutrino flux. It was shown that some of the electron neutrinos had indeed changed into the other types during their passage from the Sun to the Earth. When the total neutrino flux is measured, it is found to be in good agreement with the flux calculated from the Standard Solar model.

for billions of years. It is about 10 million times longer in fact than the estimate of a few thousand years that was obtained when the Sun was considered to be a lump of coal.

The energy has to be transported from the Sun's interior core to the surface. This is quite a slow process, and it takes about a million years for the energy to get out. The Sun's surface is cooler than the core and the energy is radiated into space as the heat and light that we observe directly. Under standard conditions, the solar power falling on the Earth is about 1.4 kilowatts per square meter ($kW\ m^{-2}$).

The first stage of the reactions just described (see also Box 3.1) where a proton converts to a neutron is known to nuclear physicists as a *weak interaction*. The process is very slow, and this sets the pace for the conversion to helium. It takes many hundreds of millions of years for two protons to fuse together. This turns out to be rather fortunate. If the fusion reaction took place

Box 3.2 The Carbon Cycle

A second chain of nuclear reactions can convert hydrogen into helium. This was proposed independently by Carl von Weizsacker and Hans Bethe in 1938, and it is now thought that it is the dominant process in stars that are hotter and more massive than the Sun. The reaction sequence is as follows:

$$p + {}^{12}C \rightarrow {}^{13}N \rightarrow {}^{13}C + e^+ + \nu$$
$$p + {}^{13}C \rightarrow {}^{14}N$$
$$p + {}^{14}N \rightarrow {}^{15}O \rightarrow {}^{15}N + e^+ + \nu$$
$$p + {}^{15}N \rightarrow {}^{12}C + {}^{4}He$$

In the first stage, a proton reacts with a ^{12}C nucleus to form nitrogen ^{13}N, which is unstable and decays to ^{13}C. Further stages build up through ^{14}N and ^{15}O to ^{15}N, which then reacts with a proton to form ^{12}C and ^{4}He. At the end of the sequence, the ^{12}C has been recycled and can start another chain of reactions, so it acts as a catalyst. Overall, four protons have been replaced with a single helium nucleus, so the energy release is the same as for the pp cycle, Figure 3.2.

too quickly, then the Sun would have burned out long before life on Earth had a chance to evolve. From our knowledge of the nuclear reaction rates and of the amount of initial hydrogen, it is estimated that the time to use up all the hydrogen is about 10 billion years. From radioactive dating of meteorites it is estimated that the age of the solar system is 4.6 billion years. Assuming the Sun is the same age as the meteorites, then it is approximately halfway through its life cycle. For comparison, the most recent estimate of the age of the universe is about 13.7 billion years.

The proton-proton reaction tends to dominate in stars that are the size of our Sun or smaller. However, in larger stars, there is another reaction cycle, involving reactions with a carbon nucleus (see Box 3.2), by which protons can be converted into helium nuclei.

3.3 Gravitational Confinement

The hydrogen in the Sun's core is compressed to very high density, roughly 10 times denser than lead. But the Sun's core is not solid—it is kept in an ionized, or *plasma*, state by the high temperature. This combination of high density and high temperature exerts an enormous outward pressure that is about 400 billion (4×10^{11}) times larger than the atmospheric pressure at the Earth's surface.

An inward force must balance this enormous outward pressure in order to prevent the Sun from expanding. Gravity provides this force in the Sun and stars, and it compresses the Sun into the most compact shape possible, a sphere. At each layer inside the sphere there has to be a balance between the outward

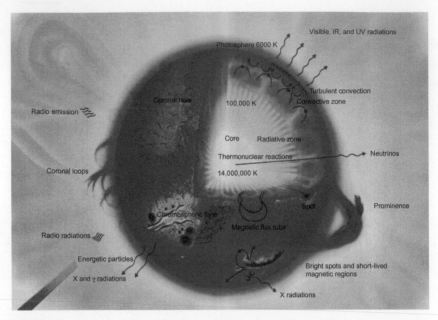

FIGURE 3.3 The main features of the Sun. Energy is released by thermonuclear reactions in the core and transported outward, first by radiation and then by convection, to the surface, from where it is radiated.

pressure and the weight of the material above (outside) pressing downward (inward). The balance between compression due to gravity and outward pressure is called *hydrostatic equilibrium*. The same effect occurs in the Earth's atmosphere: The atmospheric pressure at sea level is due to the weight of the air above—this is the combined gravitational force acting on the air molecules. The atmosphere does not collapse to a very thin layer on the ground under the pull of gravity because the upward pressure of the compressed gas in the lower layers always balances the downward pressure of the upper layers.

In some ways the structure of the Sun is similar to that of the Earth, in the sense that it has a very dense core that contains most of the Sun's mass surrounded by less dense outer layers known as the *solar envelope* (Figure 3.3). The temperature in the core is about 14 million degrees Celsius but falls quite rapidly to about 8 million degrees Celsius at a quarter of the radius and to less than 4 million degrees Celsius at half the radius. Fusion reactions are very sensitive to temperature and density and take place only in the core. The fusion power density falls to 20% of its central value at 10% of the radius and to zero outside 20% of the radius.

Fusion energy is transported outward from the core as heat, first by radiation through the layers known as the *radiative zone*. But as the radiative zone cools with increasing distance from the core, it becomes more opaque and radiation becomes less efficient. Energy then begins to move by convection through huge cells of circulating gas several hundred kilometers in diameter in the *convective zone*. Finally the energy arrives at the zone that emits the sunlight that we see, the *photosphere*. This is a comparatively thin layer, only a few hundred kilometers thick, of low-pressure gases with a temperature of 6000°C. The composition, temperature, and pressure of the photosphere are revealed by the spectrum of sunlight. In fact, helium was discovered in 1896 by William Ramsey, who found features in the solar spectrum that did not belong to any gas known on Earth at that time. The newly discovered element was named helium in honor of Helios, the mythological Greek god of the Sun.

Gravity is a very weak force compared to the forces of nuclear physics, and it can confine a hot plasma only when the mass is very large. This is possible in the Sun and stars but not for the much smaller plasmas that we would like to confine on Earth. Also, the fusion power density in the core of the Sun is very low, only 270 watts per cubic meter, compared to the megawatts per cubic meter required for a commercial power plant. Other methods of providing confinement have to be found, as is discussed in later chapters.

3.4 The Formation of Heavier Atoms

In 1929, the American astronomer Edwin Hubble, by measuring the Doppler shift of the spectral lines from many stars and galaxies, discovered that the universe is expanding. He showed that the lines are shifted to the red end of the spectrum and hence that these bodies are moving away from the Earth. The effect is often known as the *red shift*. Hubble also showed that the further the objects are away, the faster they are moving. One explanation for this effect was that there was continuous creation of matter balancing the expansion, the *steady-state theory* of the universe. However, the presently accepted theory is that everything started about 13.7 billion years ago with a gigantic explosion known as the *Big Bang*. The idea of an expanding universe was proposed in 1927 by the Belgian cosmologist Georges Lemaître, and the model was described in detail by George Gamow (Figure 3.4), Ralph Alpher, and Hans Bethe in 1948 in their famous "Alpher, Bethe, Gamow" paper. Their model predicted that there should be observable radiation left over from the Big Bang. This radiation, now known as the *Cosmic Microwave Background Radiation (CMBR)* (Box 3.3), was first observed by Arno Penzias and Robert Wilson in 1964 and was found to be close to the predicted level. The predictions of the background radiation and of the correct abundances of hydrogen, helium, and lithium, which are now observed by spectroscopy of gas clouds and old stars, are the major successes of the Big Bang theory and are the justification for taking it to be the most likely explanation for the origin of the universe.

At the inconceivably high temperatures in the primeval fireball, mass and energy were continually interchanging. As the fireball expanded, it cooled rapidly, and at this stage the energy was converted permanently into mass—first

FIGURE 3.4   George Gamow (1904–1968). Gamow was a Ukrainian, born in Odessa and educated in Leningrad, but in 1934 he emmigrated to the US, where he was a professor of physics, first at George Washington University and then at the University of Colorado. He was a prolific writer of books on science for the layperson, particularly on cosmology; many of his works are still in print.

Box 3.3 Cosmic Microwave Background Radiation

The initial observations of microwave radiation, by Penzias and Wilson at Bell Telephone Laboratories, were made quite by accident. They were testing an antenna designed for communications satellites, and in order to check the zero level of their instrument, they had pointed it at a region of the sky where they expected no radio sources. To their surprise, they obtained a radiation signal that they could not explain. A few months later, Jim Peebles at Princeton University heard of their results. He had been doing calculations based on the Big Bang theory, which predicted that the universe should be filled with a sea of radiation with a temperature less than 10 K. When they compared results, the observed radiation was in good agreement with predictions. As measurements have improved, including more sophisticated instrumentation on satellites, it was found that not only was the intensity correctly predicted, but the measured CMBR also has precisely the profile of intensity versus frequency to be consistent with the Big Bang model.

In 1992, the COBE satellite showed for the first time that there are slight variations of the CMBR intensity with direction in the sky. These observations have been even further improved using the Wilkinson Microwave Anisotropy Probe in 2002. The radiation is calculated to have been generated 380,000 years after the Big Bang—over 13 billion years ago. It shows minute variations in the temperature of the CMBR. These tiny irregularities are the seeds of the cosmic structures that have been amplified by gravitational forces to become the stars and galaxies that we see today.

as the underlying subnuclear building blocks were formed and then as these building blocks themselves combined to form protons and neutrons. Some deuterium and helium nuclei were formed, via the fusion reactions discussed earlier, when the conditions were suitable.

As the universe expanded, it became too cold for these initial fusion reactions to continue, and the mix of different nuclei that had been produced was "frozen." It is calculated that this stage was reached only 4 minutes after the initial Big Bang. At this point the universe consisted of an expanding cloud composed mainly of hydrogen (75%) and helium (25%), with small amounts of deuterium and lithium.

The universe today is known to contain 92 different elements ranging in mass from hydrogen to uranium. The theory of the Big Bang is quite explicit that nuclei much heavier than helium or lithium could not have been formed at the early stage. The obvious next question is: What is the origin of all the other elements, such as carbon, oxygen, silicon, and iron?

3.5 Stars and Supernovae

The formation of the stars occurs by gradual accretion, due to gravitational attraction, in places where there were local density variations of the material spewed out from the Big Bang. As a star forms, it is compressed by gravity, and the interior heats until it becomes sufficiently hot for fusion reactions to start heating the star still further. There is an equilibrium where the internal pressure balances the compressive force due to gravity and, when all the fuel is burned up and the fusion reaction rate decreases, gravity causes the star to contract further. Stars form in a range of different sizes and this leads to a variety of different stellar life cycles. For a star of relatively modest size like our Sun, the life cycle is expected to be about 10 billion years. When all the hydrogen has been consumed and converted into helium, the Sun will cool down and shrink in size; when the Sun is too cold for further fusion reactions, the cycle will end.

Larger stars heat to a higher temperature and therefore burn more rapidly. The fusion processes in these stars occur in a number of phases, forming more and more massive nuclei. After the first stage is completed and the hydrogen has been converted to helium, the bigger star's gravity is sufficiently strong that the star can be compressed further until the temperature rises to a value at which the helium nuclei start to fuse and form carbon in the core. This again releases energy, and the star gets even hotter. The mechanism by which helium burns was a puzzle for many years because the fusion of two helium nuclei would produce a nucleus of beryllium (^{8}Be), which is very unstable. It turns out that three helium nuclei have to join together to form a carbon nucleus (^{12}C), as explained in Box 3.4. When most of the helium has been consumed and if the star is big enough, further compression causes the temperature to rise again to the point at which the carbon burns, forming much heavier nuclei, such as neon (^{20}Ne) and magnesium (^{24}Mg). Neon is produced by the

Box 3.4 The Triple Alpha Process

When a star has converted most of its hydrogen into helium, the next stage would seem to be for two helium nuclei to combine. But this would produce ^{8}Be—a nucleus of beryllium with four protons and four neutrons that reverts back into two helium nuclei with a lifetime of less than 10^{-17} s. There is a further problem—even if the ^{8}Be nucleus is formed, the next stage in the chain,

$$^8Be + {}^4He \rightarrow {}^{12}C$$

is not allowed because the energy cannot be removed as kinetic energy with a single reaction product without violating the law of conservation of momentum. Thus, there appeared to be a bottleneck preventing the formation of the elements heavier than helium.

The English astronomer Fred Hoyle reasoned that because nuclei heavier than He do in fact exist in nature, there must be a way around the so-called *beryllium bottleneck*. He proposed that if the carbon nucleus has an excited state with energy of 7.65 MeV above the ground level of the carbon—exactly matching the energy released in the nuclear reaction—the reaction energy could be absorbed in the excited state, which could then decay to the ground state by the release of a gamma ray without any problems with momentum conservation. However, no excited state was known at the time and so Hoyle approached William Fowler at the California Institute of Technology and suggested to him that they conduct an experimental search for this state. Fowler agreed to look, and the excited state of carbon was found, thus verifying the mechanism by which the higher-mass nuclei are produced.

Overall, the triple alpha process can be looked on as an equilibrium between three ^{4}He nuclei and the excited state of ^{12}C, with occasional leakage out into the ground state of ^{12}C. This is a very slow process and is possible only in a star with enormous quantities of helium and astronomical time scales to consider.

combination of two carbon nuclei followed by the release of a helium nucleus. In succession, there are stages of neon burning and then silicon burning (Box 3.5). The reactions are shown schematically in Figure 3.5.

The detailed verification of the models of the production of all the various elements has depended very largely on many years of study of the individual nuclear processes in physics laboratories. Rather as Aston's painstaking study of the precise masses of the elements led to the eventual realization of the source of nuclear energy, so the detailed measurements of the exact types and rates of nuclear reactions under a range of different conditions enabled the detailed evolution of the universe to become understood. Measurements that were initially made in the pursuit of fundamental academic research turned out to be crucially important in the understanding of the universe. Of course, no one has made direct measurements inside the heart of a star; even the light that

Box 3.5 Heavier Nuclei

After carbon has been produced, nuclei of higher mass can be formed by reactions with further alpha particles. Each of these nuclear reactions is less efficient in terms of energy production than the previous one, because the nuclei formed are gradually becoming more stable (see Figure 2.5). The temperature increases, the reactions proceed more quickly, and the time taken to burn the remaining fuel gets shorter. In a large star, the time to burn the hydrogen might be 10 million years, while to burn the helium takes 1 million years, to burn the carbon takes only 600 years, and to burn the silicon takes less than one day! As each reaction dies down, due to the consumption of the fuel, gravity again dominates and the star is compressed. The compression continues until the star is sufficiently hot for the next reaction to start. It is necessary to reach successively higher temperatures to get the heavier elements to undergo fusion. The last fusion reactions are those that produce iron (^{56}Fe), cobalt, and nickel, the most stable of all the elements.

The principal reactions going on in various stages of the life of a massive star just prior to its explosive phase are shown in the following table (and schematically in Figure 3.5). The calculated density, temperature, and mass fraction in the various stages are shown, together with the composition in that stage.

Stage	Mass Fraction	Temp (°C)	Density (kg m^{-3})	Main Reactions	Composition
I	0.6	1×10^7	10	^{1}H$\rightarrow^4$He	^{1}H, ^{4}He
II	0.1	2×10^8	1×10^6	^{4}He$\rightarrow^{12}$C, ^{16}O	^{4}He
II	0.05	5×10^8	6×10^6	^{12}C$\rightarrow^{20}$Ne, ^{24}Mg	^{12}C, ^{16}O
IV	0.15	8×10^8	3×10^7	^{20}Ne$\rightarrow^{16}$O, ^{24}Mg	^{16}O, ^{20}Ne, ^{24}Mg
V	0.02	3×10^9	2×10^9	^{16}O$\rightarrow^{28}$Si	^{16}O, ^{24}Mg, ^{28}Si
VI	0.08	8×10^9	4×10^{12}	^{28}Si$\rightarrow^{56}$Fe, ^{56}Ni	^{28}Si, ^{32}S

we can measure remotely with telescopes and analyze by spectroscopy comes from the star's surface.

At the end of a star's lifetime, when its nuclear fuel is exhausted, the release of fusion energy no longer supports it against the inward pull of gravity. The ultimate fate of a star depends on its size. Our Sun is a relatively small star and will end its life rather benignly as a *white dwarf*, as will most stars that begin life with mass up to about two to three times that of our Sun. If the star is more massive, its core will first collapse and then undergo a gigantic explosion known as a *supernova* and in so doing will release a huge amount of energy. This will cause a blast wave that ejects much of the star's material into interstellar space. Supernovae are relatively rarely observed, typically once in every 400 years in our own galaxy, but they can also be observed in

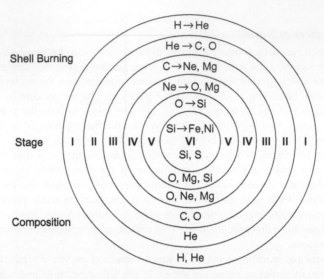

FIGURE 3.5 A schematic picture of the various fusion reactions occurring in the life of a large star, leading to the build up of more and more massive nuclei. In the final stage, the most stable elements around iron are formed. The normal chemical symbols for the elements are used. Just before the explosive stage, different reactions are occurring at different layers of the star, as shown.

other galaxies. Perhaps the most famous supernova historically is the one that was recorded by Chinese astronomers in 1054 AD. The remnants of this explosion are still observable and are known as the Crab Nebula. Since the Chinese observation, two similar explosions have taken place in our galaxy—one was observed by the famous Danish astronomer Tycho Brahe in 1572 and another by Johannes Kepler in 1604. Kepler, who was Brahe's pupil, discovered the laws of planetary motion. In 1987, the largest supernova to be observed since the invention of the telescope was seen in the Large Magellenic Cloud, the next nearest galaxy to the Milky Way. The last phase of the explosion occurred in a very short time in astronomical terms. It reached its brightest phase about 100 days after the original explosion and then started fading. Photographs taken of the original star before it exploded and then when the supernova was at its peak intensity are shown in Figure 3.6.

The importance of supernovae in forming the elements is that their temperature is very high and large numbers of energetic neutrons are produced. These are ideal conditions for production of the higher-mass elements, from iron up to uranium. The energetic neutrons are absorbed by iron and nickel to form heavier atoms. All of the elements that have been created in the stars, both during the early burning phases and during the catastrophic phase of the supernovae, are redistributed throughout the galaxy by the explosion.

An idea of how an exploding supernova disperses in the universe is seen in the photograph of the Veil Nebula, Figure 3.7. This is the remains of a supernova that exploded over 30,000 years ago. The material spreads out in fine

FIGURE 3.6 Photographs of the supernova that exploded in February 1987 in the Large Magellenic Cloud. This was the first supernova since the invention of the telescope that was bright enough to be visible to the naked eye. The view on the left shows the supernova at peak intensity, and the view on the right shows the same region before the star exploded.

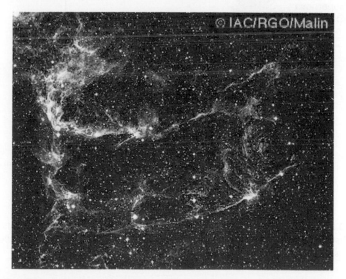

FIGURE 3.7 Photograph of the Veil Nebula showing wisps of matter remaining after the explosion of a supernova more than 30,000 years ago. The nebula is 15,000 light years away and has grown to enormous size, yet it maintains a fine filamentary structure.

wisps over an enormous volume of space. The fine material, or dust, can then gather together to form new stars and planets *and us*. The second and subsequent generations of star systems, formed from the debris of supernovae, thus contain all the stable elements. The two paths by which primary and secondary

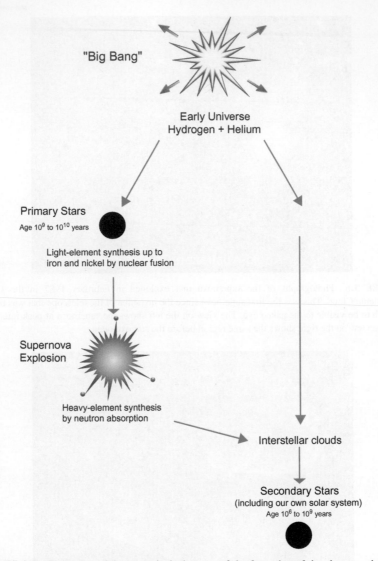

"Big Bang"

Early Universe
Hydrogen + Helium

Primary Stars
Age 10^9 to 10^{10} years

Light-element synthesis up to
iron and nickel by nuclear fusion

Supernova
Explosion

Heavy-element synthesis
by neutron absorption

Interstellar clouds

Secondary Stars
(including our own solar system)
Age 10^6 to 10^9 years

FIGURE 3.8 Illustration of the two principal stages of the formation of the elements showing how the first-generation stars contained only the light elements from the Big Bang, while second-generation stars contain the heavier elements from the supernovae explosions.

stars can form are illustrated in Figure 3.8. Our own solar system is an example of a secondary star system.

The remains of the cores of supernovae are thought to form strange objects known as *neutron stars* and *black holes*. The pull of gravity becomes so enormous that matter is squeezed tightly together, and nuclei, protons, and electrons all convert into neutrons. A neutron star has roughly one and a half times the mass of our Sun crammed in a ball about 10 kilometers in radius. Its density is therefore 100 trillion times the density of water; at that density, all the people on Earth would fit into a teaspoon! As the core gets smaller, it rotates faster and faster, like a skater who pulls his or her arms in. Strong radio radiation is emitted and can be detected on Earth as pulses, and so neutron stars are also known as *pulsars*. Neutron stars are relatively rare, only about one in a thousand stars, and the nearest one is probably at least 40 million light years away.

The stars that eventually become neutron stars are thought to start out with about 15–30 times the mass of our Sun. Stars with even higher initial masses are thought to become black holes—a region of space in which the matter forming it is crushed out of existence. The mass of a black hole is so large and the resulting gravitational field at its surface is so strong that nothing, not even light, can escape.

The roles of nuclear fusion in the universe can therefore be summarized under two main headings. First, fusion is the source of all the energy in the stars, thus supplying the energy by which we on Earth, and possibly other civilizations on the planets of other stars, survive. Second, fusion is responsible for the formation of the elements out of the primeval hydrogen. Some of the light elements (mainly hydrogen and helium) were formed by fusion in the Big Bang at the very start of the universe. The elements in the lower half of the periodic table are formed by the steady burning of the largest and hottest stars, and the very heaviest of the elements are produced by the very brief but intense reactions in the exploding supernovae.

Chapter 4

Man-Made Fusion

4.1   Down to Earth

Chapter 3 discusses the processes by which the Sun and stars release energy from fusion. There is no doubt that fusion works, but the obvious question is: Can fusion energy be made useful to mankind? The physics community had been skeptical at first about the possibility of exploiting nuclear energy on Earth; even Rutherford had gone so far as to call it "moonshine." However, speculation on the subject abounded from the days when it was suspected that nuclear processes might be important for the stars.

Impetus was added when the first atom bombs were exploded in the closing stages of World War II, with the dramatic demonstration that nuclear energy could indeed be released. If nuclear fission could release energy, why not nuclear fusion? The present chapter discusses the basic principles of how fusion energy might be exploited on Earth.

The chain of reactions in the Sun starts with the fusion of two protons—the nuclei of the common form of hydrogen—to form a nucleus of *deuterium*—the heavier form of hydrogen. When two protons fuse, one of them has to be converted into a neutron. This is the most difficult stage in the chain of reactions that power the Sun, and it takes place much too slowly to be a viable source of energy on Earth. However, after the slow first step, the fusion reactions only involve rearranging the numbers of protons and neutrons in the nucleus, and they take place much more quickly. So things look more promising if one starts with deuterium. Though deuterium is rare in the Sun, where it is burned up as fast as it is produced, on Earth there are large amounts of this form of hydrogen remaining from earlier cosmological processes. About one in every 6700 atoms of hydrogen is deuterium, and these two isotopes can be separated quite easily. The Earth has a very large amount of hydrogen, mainly as water in the oceans, so although deuterium is rather dilute, the total amount is virtually inexhaustible (Box 4.1).

The fusion reaction between two deuterium nuclei brings together two protons and two neutrons that can rearrange themselves in two alternative ways. One rearrangement produces a nucleus that has two protons and a single neutron. This is the rare form of helium known as *helium-3* (see Figure 2.2). There is a neutron left over. The alternative rearrangement produces a nucleus with one proton and two neutrons. This is the form of hydrogen known as *tritium*, which has

Fusion, Second Edition.

Box 4.1 Source of Deuterium

The reaction between deuterium and tritium has the fastest reaction rate and requires the lowest temperature of all the fusion reactions, so it is the first choice for a fusion power plant. Adequate sources of deuterium and tritium are thus important, independent of what type of confinement system, magnetic or inertial, is employed.

One gram of deuterium will produce 300 GJ of electricity, and providing for all of the world's present-day energy consumption (equivalent to about 3×10^{11} GJ per year) would require about 1000 tons of deuterium a year. The source of deuterium is straightforward because about 1 part in 6700 of water is deuterium, and 1 gallon of water used as a source of fusion fuel could produce as much energy as 300 gallons of gasoline. When all the water in the oceans is considered, this amounts to over 10^{15} tons of deuterium—enough to supply our energy requirements indefinitely. Extracting deuterium from water is straightforward using electrolysis (see Box 8.1), and the cost of the fuel would be negligible compared to the other costs of making electricity.

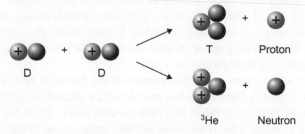

FIGURE 4.1 The two alternative branches of the fusion of two deuterium nuclei.

roughly three times the mass of ordinary hydrogen. In this case, a proton is left over. These reactions are shown schematically in Figure 4.1. Energy is released because the sum of the masses of the rearranged nuclei is slightly smaller than the mass of two deuterium nuclei, as in the fusion reactions in the Sun.

The tritium and the helium-3 produced in these reactions can also fuse with deuterium, in which case there are five nuclear particles to rearrange—two protons and three neutrons in the case of the reaction between deuterium and tritium or three protons and two neutrons in the case of deuterium plus helium-3. The result in both cases is a nucleus with two protons and two neutrons. This is the common form of helium with four units of mass—*helium-4*. It is an inert gas that is used in very-low-temperature refrigerators and can be used to fill balloons and airships. There is either a free neutron or a free proton left over. The reactions are shown schematically in Figure 4.2.

All of these reactions are used in experiments to study fusion. The reaction between deuterium and tritium, usually abbreviated as DT, requires

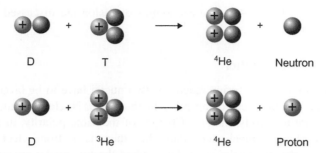

FIGURE 4.2   The reactions between deuterium and tritium or helium-3, forming helium-4.

Box 4.2 Tritium Breeding Reactions

The most convenient way to make tritium is in the reaction between neutrons and lithium. There are two possible reactions, one with each of the naturally occurring isotopes, ^{6}Li and ^{7}Li:

$$^6\text{Li} + n \rightarrow {}^4\text{He} + T + 4.8\,\text{MeV}$$
$$^7\text{Li} + n \rightarrow {}^4\text{He} + T + n - 2.5\,\text{MeV}$$

The ^{6}Li reaction is most probable with a slow neutron; it is exothermic, releasing 4.8 MeV of energy. The ^{7}Li reaction is an endothermic reaction, only occurring with a fast neutron and absorbing 2.5 MeV of energy. Natural lithium is composed of 92.6% ^{7}Li and 6.4% ^{6}Li. A kilogram of lithium will produce 1×10^5 GJ of electricity.

the lowest temperature to get it started and therefore is considered to be the best candidate for a fusion power plant. Tritium does not occur naturally on Earth because it is radioactive, decaying with a half-life of 12.3 years. This means that if we start with a fixed quantity of tritium today, only half of it will remain in 12.3 years; there will be only a quarter left after 24.6 years, and so on. Tritium has to be manufactured as a fuel. In principle, this can be done by allowing the neutron that is produced in the DT reaction to react with the element lithium (see Box 4.2). Lithium has three protons in its nucleus and exists in two forms—one with three neutrons, known as *lithium-6*, and one with four neutrons, known as *lithium-7*. Both forms interact with neutrons to produce tritium and helium. In the first reaction, energy is released, but energy has to be put into the second reaction. The basic fuels for a fusion power plant burning deuterium and tritium thus will be ordinary water and lithium. Deuterium will be extracted from water, and tritium will be produced from lithium. Both basic fuels are relatively cheap, abundant, and easily accessible. The waste product will be the inert gas helium. The overall reaction is shown schematically in Figure 4.3. The economics of fusion are discussed in more detail in Chapter 14.

4.2   Getting It Together

In order to initiate these fusion reactions, two nuclei have to be brought very close together, to distances comparable to their size. Nuclei contain protons and so they are positively charged. Charges of the same polarity, in this case two positive charges, repel each other; thus there is a strong electric force trying to keep the two nuclei apart. Only when the two nuclei are very close together does an attractive nuclear force become strong enough to counter the electric force that is trying to keep them apart. This effect is shown schematically in Figure 4.4—which plots potential energy against the distance

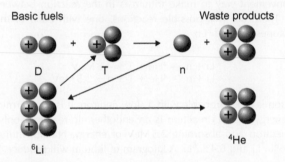

FIGURE 4.3   The overall fusion reaction. The basic fuels are deuterium and lithium; the waste product is helium.

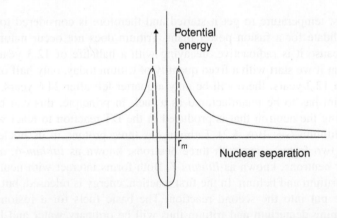

FIGURE 4.4   A diagram illustrating the potential energy of two nuclei as their distance apart is varied. When far apart, they repel each other—their electrostatic repulsion increases as they get closer together. When they become very close, a nuclear attraction becomes effective and the potential energy drops.

separating the two nuclei. Potential energy is the form of energy that a ball has on top of a hill. Bringing the two nuclei together is rather like trying to get a golf ball to roll up a hill and fall into a hole on the top. The ball has to have enough energy to climb the hill before it can fall into the hole. In this case, the hill is very much steeper and the hole much deeper and smaller than anything one would find on a golf course. Fortunately, physics comes to our aid. The laws of quantum mechanics that determine how nuclei behave at these very small distances can allow the "ball" to tunnel partway through the hill rather than having to go all the way over the top. This makes it a bit easier, but even so a lot of energy is needed to bring about an encounter close enough for fusion.

Physicists measure the energy of atomic particles in terms of the voltage through which they have to be accelerated to reach that energy. To bring about a fusion reaction requires acceleration by about 100,000 volts. The probability that a fusion reaction will take place is given in the form of a *cross-section*. This is simply a measure of the size of the hole into which the ball has to be aimed. The cross-sections for the three most probable fusion reactions are shown in Figure 4.5. In fusion golf, the effective size of the hole depends on the energy of the colliding nuclei. For DT, the cross-section is largest when the nuclei have been accelerated by about 100,000 volts (to an energy of 100 keV); it decreases again at higher energies. Figure 4.5 shows why the DT reaction is the most favorable—it offers the highest probability of fusion (the largest cross-section) at the lowest energy. Even then the hole is very small—with an area of about 10^{-28} square meters.

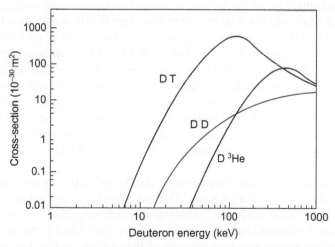

FIGURE 4.5 The probability that a fusion reaction will take place (cross-section) for a range of energies of deuterium ions. The data for three reactions are shown: deuterium plus deuterium, deuterium plus tritium, and deuterium plus helium-3. At lower energies, the probability for the DT reaction is much higher than for the other two reactions.

Voltages of hundreds of thousands of volts sound rather high compared to the hundreds of volts of a normal domestic electricity supply. However, in 1930 John Cockroft and Ernest Walton, working in Rutherford's laboratory at Cambridge University, designed and built a particle accelerator capable of generating these voltages. This sort of equipment is now commonplace in physics laboratories. In fact, physicists studying the structures within protons and neutrons use accelerators that take particles into the *gigavolt* range of energies, that is, thousands of millions of volts, and they have built *teravolt* machines, that is, millions of millions of volts. The protons and antiprotons in the Large Hadron Collider at CERN in Geneva are accelerated to energies of 7 TeV.

Accelerating nuclei to the energies needed for fusion is not difficult in the laboratory. It is relatively easy to study fusion reactions by bombarding with accelerated deuterium nuclei a solid target containing tritium. This is how the cross-sections shown earlier were measured by Marcus Oliphant and Paul Hartek in Cambridge, UK, in 1934. The problem lies with the very small cross-section of the "fusion hole" and the very steep "hill." Most of the accelerated nuclei bounce off the "hill" and never get close enough to the target nucleus to fuse. The energy that has been invested in accelerating them is lost. Only a tiny fraction of collisions (1 in 100 million) actually results in a fusion event. To return to the golfing analogy, it is rather like firing ball after ball at the hill in the hope that one will be lucky enough to find its way over the top and into the hole. Very few will make it when the par for the hole is 100 million; most balls will roll off the hill and be lost. In the case of fusion, the problem is not so much the number of lost balls but the amount of energy that is lost with them.

A better way has to be found. Clearly what is needed is a way to collect all the balls that roll off the hill and, without letting them lose energy, to send them back again and again up the slope until they finally make it into the hole. Leaving the golf course and moving indoors to the billiard or pool table illustrates how this might be done. If a ball is struck hard, it can bounce back and forth around the pool table without loosing energy (assume that the balls and the table are frictionless). Doing this with a large number of balls in motion at the same time will allow balls to scatter off each other repeatedly without losing energy. Occasionally, two balls will have the correct energies and be moving on exactly the right trajectories to allow them to fuse together when they collide. It is important to remember, however, that this happens only once in every 100 million encounters.

This picture of balls moving about randomly and colliding with each other is rather like the behavior of a gas. The gas particles—they are usually *molecules*—move about randomly, bouncing off each other and off the walls of the container, without losing any overall energy. Individual particles continually exchange energy with each other when they collide. In this way, there will always be some particles with high energies and some with low energies, but the average energy stays constant. The *temperature* of the gas is a measure of this average energy.

These considerations suggest a better way to approach fusion: take a mixture of deuterium and tritium gas and heat it to the required temperature. Known as *thermonuclear* fusion, this is to be clearly distinguished from the case where individual nuclei are accelerated and collided with each other or with a stationary target. A temperature of about 200 million degrees Celsius is necessary to give energies high enough for fusion to occur in a sufficiently high fraction of the nuclei. It is difficult to get a feel for the magnitude of such high temperatures. Remember that ice melts at 0°C, water boils at a 100°C, iron melts at around 1000°C, and everything has vaporized at 3000°C. The temperature of the core of the Sun is about 14 million degrees. For fusion reactions, it is necessary to talk in terms of hundreds of millions of degrees. To put 200 million degrees on a familiar scale would require an ordinary household thermometer about 400 kilometers long!

Collisions in the hot gas quickly knock the electrons off the atoms and produce a mixture of nuclei and electrons. The gas is said to be *ionized*, and it has a special name—it is called a *plasma* (Figure 4.6). Plasma is the fourth state of matter—solids melt to form liquids, liquids evaporate to form gases, and gases can be ionized to form plasmas. Gases exist in this ionized state in many everyday conditions, such as in fluorescent lamps and even in open flames, although in these examples only a small fraction of the gas is ionized. In interstellar space, practically all matter is in the form of fully ionized plasma, although the density

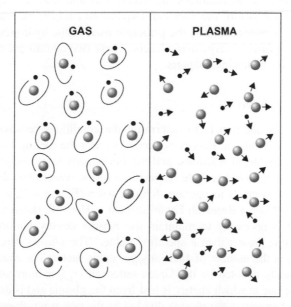

FIGURE 4.6   When a gas (shown here as an atomic gas, although many gases are molecular) is heated to high temperature, it breaks up into a mixture of negatively charged electrons and positively charged nuclei or ions.

of particles is generally very low. Since plasmas are a fundamental state of matter, their study is justified as pure science on the same basis as research into the solid, liquid, or gaseous state. At the high temperatures required for fusion, the plasma is fully ionized and consists of a mixture of negative electrons and positive nuclei. Equal numbers of negative and positive charge must be present; otherwise, the unbalanced electric forces would cause the plasma to expand rapidly. The positive nuclei are called *ions*, and this term will be used from now on. One important property of plasmas is that, with all these electrically charged particles, they can conduct electricity. The electrical conductivity of hydrogen plasma, at the temperatures required for fusion to occur, is about 10 times higher than that of copper at normal temperature.

What about the walls? The temperature of the ions has to be so high that there is no possibility of containing the hot plasma in any conventional vessel. Even the most refractory materials, such as graphite, ceramics, or tungsten, would evaporate. There are two options. One is to use a magnetic field to form a barrier between the hot fuel and the wall. The electrical charges on ions and electrons prevent them from moving directly across a magnetic field. When their motion tries to take them across the field, they simply move around in circles. They can move freely along the direction of the field, and so the overall motion is a spiral line (a helix) along the direction of the field. In this way a magnetic field can be used to guide the charged particles and prevent them from hitting the surrounding solid walls. This is called *magnetic confinement*. The second option is to compress the fusion fuel and heat it so quickly that fusion takes place before the fuel can expand and touch the walls. This is called *inertial confinement*. It is the principle used in the hydrogen bomb and when producing fusion energy using lasers. These two options are described in more detail in the following chapters.

4.3   Breaking Even

One fundamental question is to determine the conditions required for a net energy output from fusion. Energy is needed to heat the fuel up to the temperature required for fusion reactions, and the hot plasma loses energy in various ways. Clearly there would be little interest in a fusion power plant that produces less energy than it needs to operate. John Lawson (Figure 4.7), a physicist at the UK Atomic Energy Research Establishment at Harwell, showed in the mid-1950s that "it is necessary to maintain the plasma density multiplied by the confinement time greater than a specified value." The *plasma density* (usually denoted by n) is the number of fuel ions per cubic meter. The *energy confinement time*, usually denoted by the Greek letter tau (τ_E), is more subtle. It is a measure of the rate at which energy is lost from the plasma and is defined as the total amount of energy in the plasma divided by the rate at which energy is lost. It is analogous to the time constant of a house cooling down when the central heating is switched off. Of course the plasma is not allowed to cool down; the

FIGURE 4.7   John Lawson (1923–2008) explaining energy balance requirements at a meeting of the British Association in Dublin in 1957. Experimental work was still secret at this time, and the meeting was the first time that energy balance had been publicly discussed. Lawson worked in the nuclear fusion program from 1951 to 1962, but most of his career was devoted to the physics of high-energy accelerators.

objective is to keep it at a uniformly high temperature. Then the energy confinement time is a measure of the quality of the magnetic confinement. Just as the house cools down more slowly when it is well insulated, so the energy confinement time of fusion plasma is improved by good magnetic "insulation." Lawson assumed that all the fusion power was taken out as heat and converted to electricity with a specified efficiency (he took this to be about 33%, which is a typical value for a power plant). This electricity would then be used to heat the plasma. Nowadays the calculation for magnetic-confinement fusion makes slightly different assumptions but arrives at a similar conclusion.

The DT reaction produces a helium nucleus—usually known as an *alpha particle*—and a neutron. The energy released by the fusion reaction is shared between the alpha particle, with 20% of the total energy, and the neutron, with 80%. The neutron has no electric charge, and so it is not affected by the magnetic field. It escapes from the plasma and slows down in a surrounding structure, where it transfers its energy and reacts with lithium to produce tritium fuel. The fusion energy will be converted into heat and then into electricity. This is the output of the power plant. The alpha particle has a positive charge and is trapped by the magnetic field. The energy of the alpha particle can be used to heat the plasma. Initially, an external source of energy is needed to raise the plasma temperature. As the temperature rises, the fusion reaction rate increases and the alpha particles provide more and more of the required heating power. Eventually the alpha heating is sufficient by itself and the fusion reaction becomes self-sustaining. This point is called *ignition*. It is exactly analogous to using a gas torch to light a coal fire or a barbecue. The gas torch provides the external heat until the coal is at a high enough temperature that the combustion becomes self-sustaining.

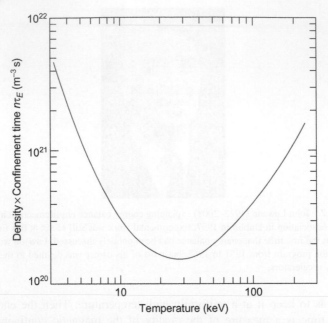

FIGURE 4.8 The ignition criterion: the value of the product of density and confinement time $n\tau_E$, necessary to obtain plasma ignition, plotted as a function of plasma temperature T. The curve has a minimum at about 30 keV (roughly 300 million degrees Celsius).

The condition for ignition in magnetic confinement is calculated by setting the alpha particle heating equal to the rate at which energy is lost from the plasma. This is slightly more stringent than the earlier version proposed by Lawson because only 20% of the fusion energy (rather than 33%) is used to heat the plasma. The ignition condition has the same form as the Lawson criterion, and the two are frequently confused. The product of density and confinement time must be larger than some specified value, which depends on the plasma temperature and has a minimum value (see Figure 4.8) in DT at about 30 keV (roughly 300 million degrees). Written in the units *particles per cubic meter × seconds*, the condition for ignition is

$$n\tau_E > 1.7 \times 10^{20} \text{ m}^{-3} \text{ s}$$

However, due to the way that the fusion cross-sections and other parameters depend on temperature, it turns out that the best route to ignition is at slightly lower temperatures. In the range 10–20 keV (100 million to 200 million degrees), the ignition condition can be written in a slightly different form that includes the temperature (see Box 4.3):

$$nT\tau_E > 3 \times 10^{21} \text{ m}^{-3} \text{ keV s}$$

Box 4.3 Conditions for Confinement

The conditions for DT magnetic-confinement fusion to reach ignition and run continuously are calculated by setting the alpha particle heating equal to the rate at which energy is lost from the plasma. Each alpha particle transfers 3.5 MeV to the plasma, and the heating power per unit volume of the plasma (in MWm^{-3}) is

$$P_\alpha = n_D n_T \overline{\sigma v} \, k \, 3.5 \times 10^3$$

The DT fusion reaction rate $\overline{\sigma v}$ ($m^3 s^{-1}$) is the cross-section σ averaged over the relative velocities v of the colliding nuclei at temperature T (keV), and n_D and n_T (m^{-3}) are the densities of D and T fuel ions. The reaction is optimum with a 50:50 fuel mixture, so $n_D = n_T = \frac{1}{2}n$, where n is the average plasma density and

$$P_\alpha = \frac{1}{4} n^2 \overline{\sigma v} \, k \, 3.5 \times 10^3 \, MWm^{-3}$$

The energy loss from the plasma is determined as follows. The average energy of a plasma particle (ion or electron) at temperature T is $(3/2)kT$ (corresponding to $\frac{1}{2}kT$ per degree of freedom). There are equal numbers of ions and electrons, so the total plasma energy per unit volume is $3nkT$. Here k is Boltzmann's constant, and when we express T in keV it is convenient to write $k = 1.6 \times 10^{-16}$ J/keV. The rate of energy loss from the plasma P_L is characterized by an energy-confinement time τ_E such that $P_L = 3nkT/\tau_E$. Setting the alpha particle heating equal to the plasma loss gives

$$n\tau_E = (12/3.5) \times 10^3 (T/\overline{\sigma v}) \, m^{-3} s$$

The right-hand side of this equation is a function only of temperature and has a minimum around $T = 30$ keV, where

$$(T/\overline{\sigma v}) \approx 5 \times 10^{22} \, KeV \, m^{-3} \, s$$

and so the minimum value of $n\tau_E$ (Figure 4.8) would be

$$n\tau_E \approx 1.7 \times 10^{20} \, m^{-3} \, s$$

Usually τ_E is also a function of temperature (see Box 10.4), and the optimum temperature comes somewhat lower than 30 keV. Fortunately, we can take advantage of a quirk of nature. In the temperature range 10–20 keV, the DT reaction rate $\overline{\sigma v}$ is proportional to T^2. Multiplying both sides of the equation for $n\tau_E$ by T makes the right-hand side $T^2/\overline{\sigma v}$, which is independent of temperature, while the left-hand side becomes the triple product

$$nT\tau_E = const \approx 3 \times 10^{21} \, m^{-3} \, keV \, s^{-1}$$

The precise value in fact depends on the profiles of plasma density and temperature and on other issues, like the plasma purity. A typical value taking these factors into account would be

$$nT\tau_E \approx 6 \times 10^{21}\,\text{m}^{-3}\,\text{keV}\,\text{s}^{-1}$$

It is important to stress that the triple product is a valid concept only for T in the range 10–20 keV.

The conditions required for a pulsed system (as in inertial-confinement fusion) can be expressed in a similar form if τ_E is defined as the pulse duration and the steady-state balance between alpha particle heating and energy loss is replaced by the assumption that all of the fusion energy is extracted after each pulse and converted into electricity, and some of the output has to be used to heat the fuel for the next pulse. The efficiency of the conversion and heating cycles for inertial confinement is discussed further in Boxes 7.1, 7.2, and 13.7.

The units are *particles per cubic meter × kilo-electron volts × seconds.* This can be expressed in different units that are a bit more meaningful for a non-specialist. The product of density and temperature is the pressure of the plasma. The ignition condition then becomes:

plasma pressure (P) × energy confinement time (τ_E) must be greater than 5.

The units are now bars multiplied by seconds, and one bar is close to the Earth's atmospheric pressure. The relationship between plasma pressure and confinement time is shown in Figure 4.9.

For magnetic-confinement fusion, an energy-confinement time of about 5 seconds and a plasma pressure of about 1 bar is one combination that could meet this

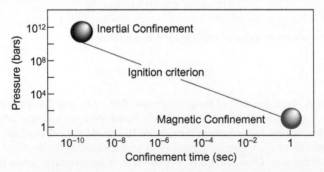

FIGURE 4.9 The conditions required for fusion plotted in terms of plasma pressure (in bars) against confinement time (in seconds). The regions for inertial-confinement and magnetic-confinement fusion are shown. In both cases, a temperature in the range 10–20 keV (roughly 100–200 million degrees Celsius) is required.

condition. It is at first sight surprising to find that the pressure of very hot plasma is close to that of the atmosphere. However, pressure is density multiplied by temperature, so a relatively low density balances the very high temperature of fusion plasma. For inertial-confinement fusion, the time is much shorter, typically less than 1 billionth (10^{-9}) of a second, and the pressure has to be correspondingly higher—more than 5 billion times atmospheric pressure. The fusion fuel has to be compressed until the density is about 1000 times higher than water.

The triple product known as $n\,T\,tau$ is used as the figure of merit against which the results of fusion experiments are compared (but note that this is valid only for temperatures in the range 10–20 keV). Progress toward ignition has required a long and difficult struggle, but now the goal is well within sight. Temperatures higher than 30 keV (300 million degrees) have been reached in some experiments, and confinement times and densities are in the right range. The most recent results in magnetic confinement, which are described in later chapters, have pushed the triple product $n\,T\,\tau_E$ up to a value that is only a factor of 5 short of ignition. The best results in inertial confinement have recently achieved values similar to those of magnetic confinement.

Chapter 5

Magnetic Confinement

5.1 The First Experiments

As with most developments in science, many people were thinking along similar lines at the same time, so it is hard to give credit to any single person as the "discoverer" of magnetic-confinement fusion. There is some evidence of speculative discussions of the possibility of generating energy from the fusion reactions before and during World War II. Certainly there had been discussions about the basic principles of fusion among the physicists building the atom bomb in Los Alamos. They had more urgent and pressing priorities at the time, and for various reasons they did not pursue their embryonic ideas on fusion immediately after the end of the war. However, many of these scientists did return later to work on fusion.

The first tangible steps were taken in the UK. In 1946, George Thomson (Figure 5.1) and Moses Blackman at Imperial College in London registered a patent for a thermonuclear power plant. Their patent was quickly classified as

FIGURE 5.1 George Thomson (on the left) and Peter Thonemann at a conference in 1972. Thomson was awarded the Nobel Prize in Physics in 1937 for demonstrating the wave characteristics of electrons.

Fusion, Second Edition.

secret, so the details were not made public at the time. In essence, the patent outlined a plan for hot plasma to be confined by a magnetic field in a dough-nut-shaped vessel that superficially looks remarkably like present-day fusion experiments. The proper geometric term for this doughnut shape—like an inflated automobile tire—is a torus. With the benefit of present knowledge of the subject, it is clear that this early idea would not have worked—but it displays remarkable insight for its day. The proposal provoked much discussion and led to the start of experimental fusion research at Imperial College.

A parallel initiative had been started in 1946 in the Clarendon Laboratory at Oxford University. Peter Thonemann (Figure 5.1) had come to Oxford from Sydney University in Australia, where, earlier in the century, the so-called *pinch effect* had been discovered. The heavy electric current that had flowed through a hollow lightning conductor during a storm had been found to have permanently squashed it. If the pinch effect was strong enough to compress metal, perhaps it could confine plasma. The basic idea is quite simple and is shown schematically in Figure 5.2. When an electric current flows through a conductor—in this case the plasma—it generates a magnetic field that encir-cles the direction of the current. If the current is sufficiently large, the mag-netic force will be strong enough to constrict, or pinch, the plasma and pull it

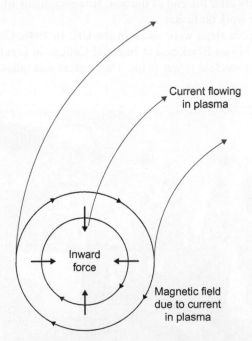

Current flowing
in plasma

Inward
force

Magnetic field
due to current
in plasma

FIGURE 5.2 Schematic drawing showing the physical mechanism that causes an electric cur-rent to compress the conductor through which it is flowing. The current in the plasma flows round the torus, producing the magnetic field. The force due to the interaction of the current and its own field is directed inward.

away from the walls. In a straight tube, the plasma will rapidly escape out of the open ends. However, if the tube is bent into a torus, it is possible in principle to create a self-constricted plasma isolated from contact with material surfaces. This is discussed in more detail in Box 5.1.

Box 5.1 Magnetic Confinement

A charged particle in a uniform magnetic field moves freely in the direction parallel to the field, but there is a force in the transverse direction that forces the particle into a circular orbit. The combined motion of the particle is a spiral, or helical, path along the direction of the magnetic field; see Figure 5.3a. The transverse radius of the helical orbit is known as the *Larmor radius* ρ (sometimes also called the *gyro radius* or *cyclotron radius*), and it depends on the charge, mass, and velocity of the particle as well as the strength of the magnetic field. The Larmor radius of an electron (in meters)

$$\rho_e = 1.07 \times 10^{-4} T_e^{0.5}/B$$

where T_e is the temperature, in kiloelectron-volts, and B is the magnetic field, in teslas. An ion with charge number Z and mass number A has a Larmor radius

$$\rho_i = 4.57 \times 10^{-3} \, (A^{0.5}/Z) T_i^{0.5}/B$$

Thus, the orbit of a deuterium ion is about 60 times larger than the orbit of an electron at the same temperature and magnetic field (see Figure 5.3).

As in a gas mixture, where the total pressure is the sum of the partial pressures of the constituents, a hot plasma exerts an outward pressure that is the sum of the kinetic pressures of the electrons and the ions; thus

$$P = n_e k T_e + n_i k T_i$$

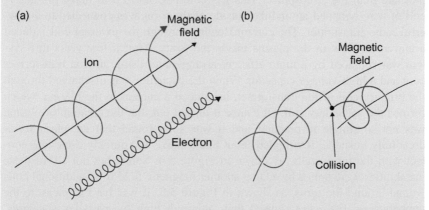

FIGURE 5.3 (a) Schematic of an ion and an electron gyrating in a straight magnetic field; (b) an ion collision resulting in the ion's being displaced to a new orbit.

where $k = 1.38 \times 10^{-23}$ J/K, or 1.6×10^{-16} J/keV, is Boltzmann's constant. For simplicity we can take $n_e = n_i$ and $T_e = T_i$, but this is not always true. In magnetic confinement, the outward pressure of the plasma has to be balanced by an inward force—and it is convenient to think of the magnetic field as exerting a pressure equal to $B^2/2\mu_0$, where B is the magnetic field strength, in teslas, and $\mu_0 = 4\pi \times 10^{-7}$ H/m is the permeability of free space. The ratio of plasma pressure to magnetic pressure is defined by the parameter $\beta = 2\,\mu_0 P/B^2$. There have been many attempts to develop magnetic-confinement configurations with $\beta \approx 1$, but the most successful routes to fusion, tokamaks and stellarators, require for stability rather low values of β, typically only a few percent.

In an ideal magnetic-confinement system, charged particles can cross the magnetic field only as a result of collisions with other particles. Collisions cause particles to be displaced from their original orbit onto new orbits (Figure 5.3b), and the characteristic radial step length is of the order of the Larmor radius. Collisions cause an individual particle to move randomly either inward or outward, but when there is a gradient in the particle density, there is a net outward diffusion of particles. The diffusion coefficient has the form ρ^2/t_c, where t_c is the characteristic time between collisions. Ions, with an orbit radius significantly larger than the electrons, would be expected to diffuse much faster than electrons, but they are prevented from doing so by the requirement that the plasma should remain neutral. A radial electric field is set up that impedes the ion cross-field diffusion rate so that it matches that of the electrons—this is known as the *ambipolar* effect. This simple classical picture fails, however, to account for the losses actually observed in magnetically confined plasmas, and we need to look to other effects (Box 10.3).

Thonemann worked with a series of small glass tori. The air inside the glass torus was pumped out and replaced with hydrogen gas at a much lower pressure than the atmosphere. This gas could be ionized to make plasma. A coil of wire wrapped around the outside of the torus was connected to a powerful radio transmitter. The current flowing through the external coil induced a current to flow in the plasma inside the torus. After a few years this system was replaced by a more efficient arrangement using an iron transformer core and a high-voltage capacitor. When the capacitor was discharged through the primary coil of the transformer, it induced a current in the plasma, which formed the secondary coil. Of course it turned out that the creation of plasma was not so simple in practice, and it was soon found that the plasma was dreadfully unstable. It wriggled about like a snake and quickly came into contact with the torus walls, as shown in Figure 5.4. Some, but not all, of these instabilities were tamed by adding another magnetic field from additional coils wound around the torus, as shown in Figure 5.5. It was hard to measure the temperatures. Estimates showed that, although high by everyday standards, they fell far short of the hundreds of millions of degrees that were required for fusion.

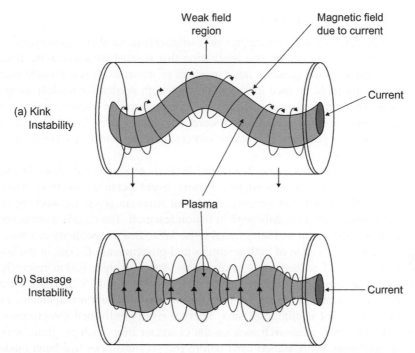

Weak field
region

Magnetic field
due to current

(a) Kink
Instability

Current

Plasma

(b) Sausage
Instability

Current

FIGURE 5.4   An illustration of some of the ways in which a plasma "wriggles" when an attempt is made to confine it in a magnetic field. If the plasma deforms a little, the outer side of the field is stretched and weakened. This leads to growth of the deformation and hence to instability.

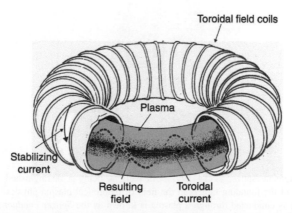

Toroidal field coils

Plasma

Stabilizing
current

Resulting
field

Toroidal
current

FIGURE 5.5   Schematic diagram of one of the first toroidal devices, the toroidal pinch. Two magnetic fields are applied: a *poloidal* field generated by the current flowing around in the plasma and a *toroidal* field produced by the external coils. The poloidal field is much stronger than the toroidal field. The combined field twists helically around the torus, as shown.

5.2   Behind Closed Doors

Fusion research looked promising, not only to scientists but also to governments. As well as producing energy, one application that seemed attractive at the time was that fusion would generate large numbers of neutrons. It was thought that these neutrons could be used to make the plutonium needed for nuclear weapons more efficiently and quickly than it could be produced in fission reactors. So fusion research was soon classified as secret and moved away from universities into more secure government research centers, like the center at Harwell near Oxford, England.

A curtain of secrecy came down, and little more was heard about fusion over the next few years. A slightly bizarre event occurred in 1951 when Argentine President Peron announced that an Austrian physicist working in Argentina had made a breakthrough in fusion research. The details were never revealed, and the claim was later found to be false. But the publicity did manage to draw the attention of both scientists and government officials in the US to the subject and so became a catalyst that activated their fusion research. An ambitious classified program was launched in 1952 and 1953. There were several experimental groups in the US that followed different arrangements of magnetic fields for confining plasma. Internal rivalry enlivened the program. Around 1954, fusion research took on the character of a crash program, with new experiments being started even before the previous ones had been made to work.

At Princeton University in New Jersey, astrophysicist Lyman Spitzer (see Figure 5.6) invented a plasma-confinement device that he called the

FIGURE 5.6 Lyman Spitzer (1914–1997). As well as being a distinguished astrophysicist, Spitzer was one of the founding fathers in the field of theoretical plasma physics. The ease with which electricity is conducted through a plasma is known as the *Spitzer conductivity*. He was a professor of astronomy at Princeton University from 1947 to 1979, and, among other things, he is known for being the first person to propose having a telescope on a satellite in space. His foresight is recognized by the naming of the Spitzer Space Telescope, which was launched into space by a Delta rocket from Cape Canaveral, Florida, on August 25, 2003.

stellarator. Unlike the pinch, where the magnetic field was generated mainly by currents flowing in the plasma itself, the magnetic field in the stellarator was produced entirely by external coils. In a pinch experiment, the plasma current flows around inside the torus—this is called the *toroidal* direction—and generates a magnetic field wrapped around the plasma in what is called the *poloidal* direction (see Figure 5.5 and Box 5.2). The original idea for the stellarator had been to confine the plasma in a toroidal magnetic field. It was quickly realized that such a purely toroidal magnetic field cannot confine plasma and that it was necessary to add a twist to the field. In the first

---

Box 5.2 Toroidal Confinement

The earliest magnetic-confinement devices were developed in the UK in the late 1940s. These were *toroidal pinches* (Figure 5.5), which attempted to confine plasma with a strong, purely poloidal magnetic field produced by a toroidal plasma current. With a sufficiently strong current, the magnetic field compresses (or *pinches*—hence the name) the plasma, pulling it away from the walls. But this arrangement proved seriously unstable—the plasma thrashed about like a snake or constricted itself like a string of sausages (Figure 5.4). External coils adding a weak magnetic field in the toroidal direction improved stability, and further improvement was found when this toroidal field reversed direction outside the plasma—a configuration now known as the *reverse field pinch* (RFP). Initially, the field reversal occurred spontaneously, but nowadays it is induced deliberately. The potential to work at high β (see Box 5.1) would be a possible advantage of the RFP, but good energy confinement has proved elusive.

The second approach to toroidal confinement is the *stellarator*, invented at Princeton in the early 1950s. The stellarator evolved as an attempt to confine fusion plasmas by means of a strong toroidal magnetic field produced by an external toroidal solenoid—without any currents in the plasma. But such a purely toroidal field cannot provide the balancing force against expansion of the plasma (this is one of the reasons the toroidal theta pinch failed). It is necessary to twist the magnetic field as it passes around the torus so that each field line is wrapped around the inside as well as the outside of the cross-section (see Figure 5.7). The coils in modern stellarators have evolved in various ways but share the same basic principle of providing a twisted toroidal magnetic field. Stellarators fell behind tokamaks in the 1960s and 1970s but now claim comparable scaling of confinement time with size. The largest experiments are the *Large Helical Device* (LHD), which came into operation at Toki in Japan in 1998, and the W7-X machine, under construction at Greifswald in Germany (see section 10.8).

The third, and most successful, toroidal confinement scheme is the *tokamak*, developed in Moscow in the 1950s. The tokamak can be thought of either as a toroidal pinch with a very strong stabilizing toroidal field or as using the poloidal field of a current in the plasma to add the twist to a toroidal field. It is now the leading contender for a magnetically confined fusion power plant. The tokamak is described in more detail in Chapter 9.

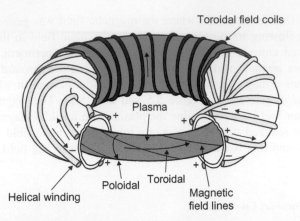

FIGURE 5.7   Schematic of a stellarator. The outer winding provides the toroidal field, and the inner helical winding provides the poloidal field that gives the field lines a twist, causing the magnetic field lines to spiral around inside the chamber. The toroidal field is much stronger than the poloidal field.

experiments this was done simply by physically twisting the whole torus into the shape of a "figure 8." Later, the same effect was produced using a second set of twisted coils, the *helical winding* shown in Figure 5.7. The stellarator has the advantage that it is capable of operating continuously. It is still considered to have the potential to be the confinement system for a fusion power plant, and active research on stellarators is being pursued in some fusion laboratories.

The fusion research program at Los Alamos in New Mexico studied toroidal pinches similar to those in the UK and also *linear theta pinches*, where a strong magnetic field is built up very rapidly to compress the plasma. It was thought that if this could be done sufficiently quickly, fusion temperatures might be reached before the plasma had time to escape out of the open ends. The Lawrence Livermore National Laboratory in California also built some pinches but concentrated on confining plasma in a straight magnetic field by making it stronger at the ends—the so-called *mirror machine*. A similar approach using mirror machines was followed at Oak Ridge in Tennessee. Some of these linear magnetic configurations are sketched in Figure 5.8 and outlined in Box 5.3. Although a linear system would have advantages compared to a toroidal one, in terms of being easier to build and maintain, there are obvious problems with losses from the ends. A full discussion of all the alternatives that were explored is outside the scope of this book, which will concentrate on the most successful lines that were followed.

In the Soviet Union, the first proposal for a fusion device came in 1950 from Oleg Lavrentiev. He was a young soldier, without even a high school diploma, who was serving in the Soviet army. His proposal to confine plasma with electric rather than magnetic fields was passed on to scientists in Moscow. They

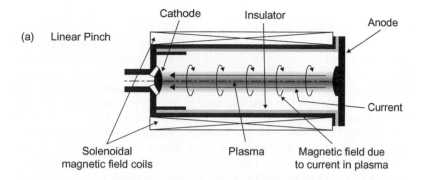

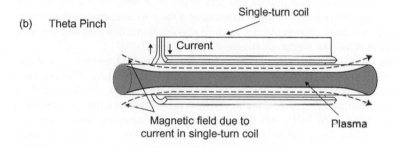

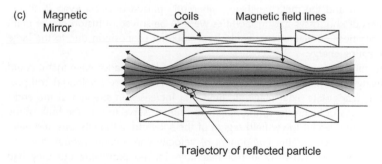

FIGURE 5.8 Sketch of three different linear magnetic configurations that have been considered for the confinement of hot plasma. All the linear systems had excessive losses through the ends and were abandoned.

concluded that electrostatic confinement would not work in this particular way, but they were stimulated to pursue magnetic confinement. This led to a strong program, initially on pinches but also expanding into the other areas, such as the open-ended mirror machines and later into the tokamak, as discussed in Chapter 9. Lavrentiev's story is remarkable. He was called to Moscow, where he finished his education with personal tuition from leading professors and his career in fusion research continued at Kharkov in the Ukraine.

Box 5.3 Linear Confinement

In the *linear Z-pinch*, a plasma current flowing along the axis between two end electrodes produces an *azimuthal* magnetic field that compresses (or pinches) the plasma away from the walls. Unfortunately, the plasma rapidly becomes unstable, breaks up, and hits the wall. In fact, many linear Z-pinch experiments were built to study the instabilities that had been seen in toroidal pinches. A variant of the Z-pinch, known as the *plasma focus*, was studied intensively during the 1960s. It has long since been abandoned as a serious candidate for a fusion power plant, but it continues to provide a source of dense plasma for academic studies. The linear Z-pinch survives today in the form of ultrafast, very-high-current devices (sometimes using metal filaments to initiate the plasma), where it is hoped that the plasma compression and heating can take place on a faster time scale than the instabilities. The potential of these schemes as a route to magnetic confinement is doubtful, but these ultrafast pinches produce copious bursts of X-rays and are being studied as drivers for inertial-confinement fusion (Chapter 7).

The *theta pinch* is generated by a fast-rising azimuthal current in a single-turn external conductor that is wrapped around the plasma tube. This produces an axial magnetic field that compresses and heats the plasma. The SCYLLA theta pinch, developed at Los Alamos in 1958, was the first magnetic-confinement system to produce really hot fusion plasmas with neutrons from thermonuclear reactions. The pulse duration of a theta pinch is very short, typically about 1 microsecond, but even on this short time scale plasma is lost by instabilities and end losses. Attempts to stopper the ends with magnetic fields or material plugs and by joining the two ends to make a toroidal theta pinch all failed. Both the Z-pinch and the theta pinch are inherently pulsed devices. Even if all the problems of end losses and instabilities were to be solved, a fusion power plant based on these configurations would be a doubtful proposition due to the large re-circulating energy.

The third linear magnetic-confinement scheme has the advantage that it could run steady state. This is the *magnetic mirror machine*, where a solenoid coil produces a steady-state axial magnetic field that increases in strength at the ends. These regions of higher field, the magnetic mirrors, serve to trap the bulk of the plasma in the central lower field region of the solenoid, although ions and electrons with a large parallel component of velocity can escape through the mirrors. At low plasma density, mirror machines looked promising but they had difficulties in reaching the higher densities needed for a power plant. Instabilities and collective effects caused losses that could not be overcome, in spite of adding complicated end cells to the basic mirror configuration. Development of mirror machines in the US was stopped in 1986 and programs in the former Soviet Union have been wound down due to lack of research funding. A mirror machine, GAMMA 10, still operates in Tsukuba, Japan.

5.3   Opening the Doors

All this research was conducted in great secrecy. Scientists in the UK and the US knew something of each other's work, but not the details. Very little was known about the work in the Soviet Union. Should fusion research be

FIGURE 5.9   The visit of the Soviet delegation to AERE Harwell in 1956. John Cockroft (the Director of Harwell) is in the left foreground and next to him is Igor Kurchatov (1903–1960). Nikita Kruschev is hidden behind Kurchatov. Nikolai Bulganin is on Kruschev's left.

continued in secrecy or be made open? The pressure for declassification was increased in a dramatic way. In 1956, the Soviet leaders Nikita Khrushchev and Nikolai Bulganin went to the UK on a highly publicized state visit. With them was the distinguished physicist Igor Kurchatov (Figure 5.9), who visited the Atomic Energy Research Establishment at Harwell and offered to give a lecture "On the Possibility of Producing Thermonuclear Reactions in a Gas Discharge." This was a surprise and of great interest to the British scientists working on the "secret" fusion program. It was difficult for them to ask questions without revealing what they knew themselves, and for that they would have needed prior approval. They discovered that Soviet scientists had been following very similar lines of research into magnetic confinement as the UK and the US, concentrating on both straight and toroidal pinch experiments. Kurchatov gave an elegant lecture outlining some of the main Soviet discoveries. He warned that it is possible to get production of neutrons, indicating that fusion reactions are occurring, without having a true thermonuclear reaction.

Kurchatov was one of the stars of the Soviet scientific establishment. He had supervised the building of the Soviet atom bomb, and in 1949 he laid down plans for the first Soviet nuclear (fission) power plant. In 1951 he organized the first Soviet conference on controlled thermonuclear fusion, and a few months later he had set up a laboratory headed by Lev Artsimovich. Kurchatov, who died in 1960, was an advocate of nuclear disarmament; he recognized the importance of having open fusion research, and his lecture at Harwell was his first attempt at achieving it. The first open presentation of work on fusion research was at an international conference in Geneva in 1958.

5.4 ZETA

Fusion research at Harwell had expanded rapidly through the early 1950s. Bigger and ever more powerful pinch experiments were built, culminating in the ZETA machine, which started operation in 1957. ZETA was a bold venture and a remarkable feat of engineering for its day (Figure 5.10). Its aluminum torus of 3 meters diameter and 1 meter bore was originally intended to contain a plasma current of 100,000 amps, but this specification was soon raised to 900,000 amps. Already in its first few weeks of operation in deuterium plasmas, currents were running at nearly 200,000 amps, and large numbers of neutrons were recorded, up to a million per pulse. This caused great excitement, but the important question was, were these neutrons *thermonuclear*?

FIGURE 5.10 The ZETA toroidal pinch experiment at Harwell soon after its first operation in 1957. The large transformer core (on the right-hand side of the photograph) surrounds the toroidal vacuum vessel.

Kurchatov had already warned of the possibility that beams of ions might be accelerated to high energies and produce neutrons that could be misinterpreted as coming from the hot plasma. The distinction, though subtle, was very important and would seriously affect the way the results would extrapolate to a fusion power plant. The uncertainty could have been resolved if it had been possible to measure the plasma temperature accurately, but the techniques to do this were still in their infancy.

News of the existence of the top-secret ZETA machine and of its neutrons quickly leaked to the press. Pressure grew for an official statement from Harwell, and delays in doing so merely heightened the speculation. It was decided to publish the results from ZETA alongside papers from other British and American fusion experiments in the scientific journal *Nature* at the end of January 1958. The carefully worded ZETA paper made no claims at all about the possible thermonuclear origin of the neutrons. Harwell's director, John Cockcroft, was less cautious, and at a press conference he was drawn into saying that he was "90% certain" that the neutrons were thermonuclear. The press reported this with great enthusiasm, publishing stories about cheap electricity from seawater. The matter was soon resolved by a series of elegant measurements that showed the neutrons were not thermonuclear. ZETA went on to make many valuable contributions to the understanding of fusion plasmas, but plans to build an even larger version were abandoned.

5.5 From Geneva to Novosibirsk

A few months after the publication of the ZETA results, the final veil of secrecy was lifted from fusion research at the Atoms for Peace conference held by the United Nations in Geneva in September 1958. Some of the actual fusion experiments were taken to Geneva and were shown working at an exhibition alongside the conference. This was the first chance for fusion scientists from communist and capitalist countries to see and hear the details of each other's work, to compare notes, and to ask questions. In many cases they found that they had been working along similar lines and had made essentially the same discoveries quite independently. Above all, the conference provided an opportunity for the scientists to meet each other. Personal contacts were established that would lead to a strong tradition of international collaboration in future fusion research.

The many different magnetic configurations that had been tested could be grouped into two main categories according to whether the magnetic field was open at the ends (like the theta and Z pinches and the mirror machines) or closed into a torus (like the toroidal pinches and stellarators). Each of these categories could be further divided into systems that were rapidly pulsed and systems with the potential to be steady state. In order to meet the *Lawson criterion* for fusion, which set a minimum value for the product of density and confinement time (Box 4.3), rapidly pulsed systems aiming to contain plasma for at most a few thousandths of a second clearly had to work at higher densities than systems that aimed to contain the plasma energy for several seconds.

Leading the field in the late 1950s in terms of high plasma temperatures and neutron production were some of the rapidly pulsed open-ended systems—in particular the theta pinches. These had impressive results because the plasma was heated by the rapid initial compression of the magnetic fields before it had time to escape or become unstable, but their potential as power plants was limited by loss of plasma through the open ends and by instabilities that developed relatively quickly. The open-ended mirror machines also suffered from end losses. The toroidal pinches and stellarators avoided end losses, but it was found that plasma escaped across the magnetic field much faster than expected. If these prevailing loss rates were scaled up to a power plant, it began to look as if it might be very difficult—perhaps even impossible—to harness fusion energy for commercial use.

In the 1960s, progress in fusion research seemed slow and painful. The optimism of earlier years was replaced by the realization of just how difficult it was to use magnetic fields to contain hot plasma. Hot plasmas could be violently unstable and, even when the worst instabilities were avoided or suppressed, the plasma cooled far too quickly. New theories predicted ever more threatening instabilities and loss processes. Novel configurations to circumvent the problems were tried, but they failed to live up to their promise. Increasingly, emphasis shifted from trying to achieve a quick breakthrough into developing a better understanding of the general properties of magnetized plasma by conducting more careful experiments with improved measurements.

Attempts to reduce the end losses from the open systems met with little success, and one by one these lines were abandoned. Toroidal systems, the stellarators and pinches, made steady progress and slowly began to show signs of improved confinement, but they were overtaken by a new contender—the *tokamak*—a configuration that had been invented in Moscow at the Kurchatov Institute. The name is an acronym of the Russian words *toroidalnaya kamera*, for "toroidal chamber," and *magnitnaya katushka*, for "magnetic coil." In 1968, just 10 years after the conference in Geneva, another major conference on fusion research was held—this time in the Siberian town of Novosibirsk. The latest results presented from the Russian tokamaks were so impressive that most countries soon decided to switch their efforts into this line. This story is continued in Chapter 9.

Chapter 6

The Hydrogen Bomb

6.1 The Background

Some isotopes of uranium and plutonium have nuclei that are so close to being unstable that they fragment and release energy when bombarded with neutrons. A fission chain reaction builds up because each fragmenting nucleus produces several neutrons that can initiate further reactions. An explosion occurs if the piece of uranium or plutonium exceeds a certain *critical mass*— thought to be a few kilograms (i.e., smaller than a grapefruit). In order to bring about the explosion, this critical mass has to be assembled very quickly, either by firing together two subcritical pieces or by compressing a subcritical sphere using conventional explosives. The US developed the first atom bombs in great secrecy during World War II at Los Alamos, New Mexico. The first test weapon, exploded in New Mexico in July 1945, had a force equivalent to 21 kilotons of high explosive. A few days later, bombs of similar size devastated the Japanese cities of Hiroshima and Nagasaki.

Producing the fissile materials for such weapons was difficult and expensive and required an enormous industrial complex. Less than 1% of natural uranium is the "explosive" isotope ^{235}U, and separating it from the more abundant ^{238}U is very difficult. Plutonium does not occur naturally at all and must be manufactured in a fission reactor and then extracted from the intensely radioactive waste. Moreover, the size of a pure fission bomb was limited by the requirement that the component parts be below the critical mass. Fusion does not suffer from this size limitation and might allow bigger bombs to be built. The fusion fuel, deuterium, is much more abundant than ^{235}U and is easier to separate.

Even as early as 1941, before he had built the very first nuclear (fission) reactor in Chicago, the physicist Enrico Fermi speculated to Edward Teller that a fission bomb might be able to ignite the fusion reaction in deuterium in order to produce an even more powerful weapon—this became known as the *hydrogen bomb*, or *H-bomb*. These ideas were not pursued seriously until after the war ended. Many of the scientists working at Los Alamos then left to go back to their academic pursuits. Robert Oppenheimer, who had led the development of the fission bomb, resigned as director of the weapons laboratory at Los Alamos to become director of the Princeton Institute of Advanced Study and was replaced by Norris Bradbury. Edward Teller (Figure 6.1), after a brief period in academic life, returned to Los Alamos and became the main driving

FIGURE 6.1   Edward Teller (1908–2003). Born in Budapest, he immigrated to the US in 1935. As well as proposing the hydrogen bomb, he was a powerful advocate of the "Star Wars" military program.

force behind the development of the H-bomb, with a concept that was called the *Classical Super*.

There was, however, much soul-searching in the US as to whether it was justified to try and build a fusion bomb at all. In 1949, Enrico Fermi and Isidor Rabi, both distinguished physicists and Nobel Prize winners, wrote a report for the Atomic Energy Commission in which they said:

Necessarily such a weapon goes far beyond any military objective and enters the range of very great natural catastrophes.... It is clear that the use of such a weapon cannot be justified on any ethical ground which gives a human being a certain individuality and dignity even if he happens to be a resident of an enemy country.... The fact that no limits exist to the destructiveness of this weapon makes its existence and the knowledge of its construction a danger to humanity as a whole. It is an evil thing considered in any light.

The debate was cut short in early 1950 by the unexpected detonation of the first Soviet fission bomb. Prompted by the suspicion that East German spy Klaus Fuchs had supplied information about US hydrogen-bomb research to the Soviet Union, President Truman ordered that the Super be developed as quickly as possible. However, no one really knew how to do this, so new fears were raised that Truman's statement might simply encourage the Soviet Union to speed up its own efforts to build a fusion bomb and, more seriously, that the Soviet scientists might already know how to do it.

6.2   The Problems

It was clear that developing a fusion weapon would be even more difficult than development of the fission bomb had been. It was necessary to heat the fusion fuel very quickly to extremely high temperatures to start the fusion reaction and to obtain the conditions for the reaction to propagate like a flame throughout all the fuel. It was recognized that the only way to do this was to set off a fission bomb and to use the tremendous burst of energy to ignite the fusion fuel. At first it was hoped that the fission bomb simply could be set off adjacent to the fusion fuel and that the heat would trigger the fusion reactions. However, calculations showed that this approach was unlikely to work. The shock wave from the fission bomb would blow the fusion fuel away long before it could be heated to a high-enough temperature to react. The problem has been likened to trying to light a cigarette with a match in a howling gale.

There were further serious problems in terms of providing the fusion fuel. A mixture of deuterium and tritium would ignite most easily, but tritium does not occur naturally and must be manufactured in a nuclear reactor. A DT fusion bomb with an explosive yield equal to that of 10 million tons of TNT (10 megatons) would require hundreds of kilograms of tritium. The rough size of the bomb can be estimated from the density of solid deuterium; it would be equivalent to a sphere about 1 meter in diameter. Smaller quantities of deuterium and tritium could be used to boost the explosive force of fission bombs or to initiate the ignition of a fusion weapon. But to manufacture tritium in the quantities needed for a significant number of pure DT bombs would require a massive production effort—dwarfing even the substantial program already under way to manufacture plutonium—and would be prohibitively expensive. Attention focused therefore on the even more difficult task of igniting explosions in pure deuterium or in a mixture of deuterium and lithium. The lithium could be converted into tritium *in situ* using neutrons that would be produced as the fuel burned.

Teller's original idea had been to explode a small fission bomb at one end of a long cylinder of frozen deuterium. The basic idea was that the fission bomb would heat the fusion fuel sufficiently to ignite a fusion reaction in the deuterium. A refinement was to use a small amount of tritium mixed with the deuterium to help start the ignition. If this worked, in principle there would be no limit to the strength of the explosion, which could be increased just by making the deuterium cylinder longer. However, calculations by Stanislaw Ulam, another Livermore scientist, had cast doubts on its feasibility and had shown that unrealistically large quantities of tritium would be required. Teller proposed a number of new designs in late 1950, but none seemed to show much promise.

Early in 1951, Ulam made an important conceptual breakthrough, and Teller quickly refined the idea. This followed an idea known as radiation implosion that had first been broached in 1946 by Klaus Fuchs before he was arrested for giving atomic secrets to the Soviet Union. Most of the energy leaves the fission trigger as X-rays. Traveling at the speed of light, the X-rays can reach the nearby fusion fuel almost instantaneously and be used to compress and ignite it before it is blown apart by the blast wave from the fission

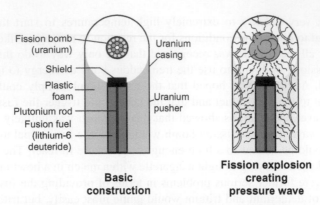

FIGURE 6.2   Schematic diagram of the elements of an H-bomb. A fission explosion is first triggered by high explosive. This explosion is contained inside a heavy metal case. The radiation from the fission bomb causes the implosion and heating of the fusion fuel and sets off the fusion bomb.

explosion, which travels only at the speed of sound. This is analogous to the delay between seeing a flash of lightning and hearing the sound of the thunder, though the much smaller distances in the bomb reduce the delay to less than a millionth of a second. A second important requirement is that the radiation from the fission bomb needs to compress the fusion fuel before it is heated to the high temperature at which it will ignite. It is much easier to compress a gas when it is cold than when it is hot.

The technique for compressing the fusion fuel has since become known as the *Teller-Ulam configuration*, and it is shown schematically in Figure 6.2. Although many of the details remain secret, in 1989 there was a partial declassification of the history of American hydrogen bomb development, and the basic principles are known. A description of the bomb is now included in the *Encyclopedia Britannica*. Like all technical achievements, once it is known that it can be done, it is much easier to work out how to do it.

The fission bomb trigger is set off at one end of a cylindrical casing in which the fusion fuel is also contained. The fusion fuel is thought to be in the form of a cylinder surrounding a rod of plutonium, and a layer of very dense material—usually natural uranium or tungsten—surrounds the fuel itself. The X-ray radiation from the fission bomb is channeled down the radial gap between the outer casing and the fusion fuel cylinder. The gap is filled with plastic foam that is immediately vaporized and turned into hot plasma. The plasma is transparent to the X-rays, allowing the inner surface of the cylindrical casing and the outer surface of the dense layer surrounding the fusion fuel to be heated quickly to very high temperatures. As the outer surface of the layer surrounding the fuel vaporizes, it exerts a high inward pressure—rather like an inverted rocket engine.

Enormous pressures are generated instantaneously—several billion times atmospheric pressure—and the fuel is compressed to typically 300 times its

FIGURE 6.3   Photograph of the explosion of the first fusion bomb over Eniwetok Atoll in November 1952.

normal density. It is salutary to realize that the explosive force released by the fission trigger, enough to destroy an entire city, is being used briefly to squeeze a few kilograms of fuel! The compression and the neutrons from the fission bomb cause the plutonium rod down the middle of the fusion fuel to become critical and to explode—in effect a second fission bomb goes off. This explosion rapidly heats the already compressed fusion fuel to the temperature required for the fusion reactions to start. Once ignited, the fusion fuel burns outward and extremely high temperatures—up to 300 million degrees—are reached as almost the whole of the fuel is consumed. Some reports suggest that the design has been refined so much that the plutonium "spark plug" is no longer needed.

The first thermonuclear bomb test, code named Ivy-Mike, took place at Eniwetok Atoll in the Pacific Ocean in November 1952 and achieved a yield equivalent to 10 megatons (Figure 6.3). It is estimated that only about a quarter of this came from fusion and the rest from fission induced in the heavy uranium case. This was a test of the principle of the compression scheme rather than a deployable weapon. The fusion fuel was liquid deuterium, which had to be refrigerated and contained in a large vacuum flask. A massive casing of uranium and steel surrounded this. The whole device weighed over 80 tons—hardly a weapon that could be flown in an airplane or fired from a missile.

In order to make a more compact weapon, a version using solid fusion fuel, lithium deuteride, was developed. In this case, tritium is produced in the course of the reaction by neutron bombardment of the lithium, and the principal fusion reaction is between deuterium and tritium. Lithium deuteride, a chemical compound of lithium and deuterium, is a solid at room temperature, and this approach obviates the necessity of the complex refrigeration system. The US tested this concept in the spring of 1954, with a yield of 15 megatons.

FIGURE 6.4   Andrei Sakharov (1921–1989). He made many contributions to the nuclear fusion program, and, along with Igor Tamm, he is credited with the invention of the tokamak. He played a crucial role in the Soviet atomic weapons program until 1968, when he published his famous pamphlet, *Progress, Peaceful Coexistence and Intellectual Freedom*. He was awarded the Nobel Peace Prize in 1975 but was harassed by the Soviet authorities for his dissident views.

6.3   Beyond the "Sloyka"

The Soviet Union had also started to think about fusion weapons in the late 1940s and recognized the same difficulties as the Americans in igniting explosions in pure deuterium. The first Soviet H-bomb test, in 1953, consisted of alternating layers of lithium deuteride and uranium wrapped around the core of a fission bomb. This concept, by Andrei Sakharov (Figure 6.4), was called the sloyka—a type of Russian layer cake. The principle was to squeeze the fusion fuel between the exploding core and the heavy outer layers. Teller had apparently considered something rather similar that he called the "alarm clock." This was not a true H-bomb, in that most of the energy came from fission, and its size was limited to less than a megaton.

Andrei Sakharov is also credited with independently conceiving the idea of a staged radiation implosion, similar to that of Teller and Ulam, which led to the first true Soviet H-bomb test in November 1955. It is thought that most subsequent thermonuclear weapons, including those developed by other countries, have been based on principles similar to the Teller-Ulam configuration that has been described here, although of course most of the details remain secret.

6.4   Peaceful Uses?

Chemical explosives, such as TNT and dynamite, are used extensively in mining and construction—and both the United States and the Soviet Union have tried hard to find similar peaceful uses for nuclear explosions. The US program—which was known as Project Plowshare—took its name from the biblical quotation "and they shall beat their swords into plowshares." The project considered civil engineering applications, such as using a series of nuclear explosions to either widen the existing Panama Canal or to construct a new sea-level waterway connecting the Atlantic and Pacific oceans, to construct new dams and harbors and to cut railroads and highways through mountainous areas. Other ideas involved blasting underground caverns for storing water, natural gas, or petroleum and using underground explosions to improve the flow from difficult oil or natural gas fields by fragmenting rock formations that have low natural permeability. The US carried out nearly thirty underground explosions during the 1960s and '70s to study some of these ideas, but Project Plowshare was ended in 1977 due to growing concerns about radioactive fallout and about the ethics of using nuclear weapons under any circumstances.

In the same era, the Soviet Union had an even larger program that carried out over one hundred nuclear explosions to explore their possible peaceful uses. Several applications, such as deep seismic sounding and oil stimulation, were explored in depth and appeared to have a positive cost-benefit ratio at minimal public risk, but other applications developed significant technical problems—and sometimes things did go wrong, casting a shadow on their general applicability. Some specialized applications, such as using a nuclear explosion to seal a runaway oil or gas well, demonstrated a unique technology that one day may have to be considered for use as a last resort to avoid an even greater environmental disaster. Other ideas were the subject of one or two exploratory tests but were not pursued further for reasons that have never been explained. Overall, the Soviet program for peaceful uses represented a significant technical effort to explore what was seen at the time to be a promising new technology—and it generated a large body of data, although only a small fraction of this has been made public.

One peaceful application studied in the United States was the possibility of converting the energy released by a nuclear explosion into electricity. The simplest idea was to set off a small nuclear bomb deep underground inside a natural formation of rock salt. The energy of the explosion would melt the salt—and then water would be pumped into the cavity and the steam extracted to generate electricity. This concept was tested in December 1961 in the first explosion of the Plowshare series—a test known as *Gnome* using a small fission bomb with a yield of about 3 kilotons.

The concept was followed up in the mid-1970s at Los Alamos National Laboratory with a study (known as the *Pacer Project*) of a more sophisticated scheme to generate a steady supply of electricity from nuclear explosions. A typical design envisaged exploding the bombs inside an underground blast-chamber—a steel cylinder 30 m in diameter and 100 m tall with 4 m thick walls filled with molten fluoride salt. The molten salt would absorb the energy of the explosions (and also the neutrons, to reduce damage to the blast-chamber) and would then be used to heat water to drive a steam turbine. A 1 kiloton bomb releases about 4000 GJ (the same amount of energy that is obtained by burning about 175 tons of coal) and, at the planned rate of one bomb every 45 min, the power output (if converted to electricity with the 30% efficiency typical of a coal-fired steam plant) would be about 500 MW. However, a large supply of nuclear bombs, roughly 10,000 bombs per year, would be required for the plant to operate continuously. Leaving aside the environmental and political issues involved in producing nuclear bombs on such a massive scale, the economics of such a system are very doubtful given the enormous costs and difficulties of producing fissile material. In principle, it would be more economical to use fusion bombs, which deliver much more energy per unit of fissile material, but in practice it would be impossible to contain explosions larger than a few kilotons within a realistic engineered structure. It is not surprising that Project Pacer never progressed beyond the conceptual stage.

In the next chapter we look at the prospects for producing controlled fusion energy by exploding miniature fusion "bombs" in rapid succession without the need for a fission trigger—if this can be achieved, it would offer an alternative to magnetic-confinement fusion.

Chapter 7

Inertial-Confinement Fusion

7.1 Mini-Explosions

The inertial-confinement route to controlled-fusion energy is based on the same general principle as that used in the hydrogen bomb—fuel is compressed and heated so quickly that it reaches the conditions for fusion and burns before it has time to escape. The inertia of the fuel keeps it from escaping—hence the name inertial-confinement fusion (ICF).

Of course, the quantity of fuel has to be much smaller than that used in a bomb, so that the energy released in each "explosion" will not destroy the surrounding environment. The quantity of fuel is constrained also by the amount of energy needed to heat it sufficiently quickly. These considerations lead to typical values of the energy that would be produced by each mini-explosion that lie in the range of several hundred million joules. To put this into a more familiar context, 1 kilogram of gasoline has an energy content of about 40 million joules, so each mini-explosion would be equivalent to burning about 10 kilograms of gasoline. Due to the much higher energy content of fusion fuels, this amount of fusion energy can be produced by burning a few milligrams of a mixture of deuterium and tritium—in solid form this is a small spherical pellet, or capsule, with a radius of a few millimeters. An inertial fusion power plant would have a chamber where these mini-explosions would take place repeatedly in order to produce a steady output of energy. In some ways this would be rather like an automobile engine powered by mini-explosions of gasoline. The main sequence of events is shown schematically in Figure 7.1.

The general conditions for releasing energy from fusion discussed in Chapter 4 are essentially the same for inertial confinement as for magnetic confinement. To recap, for fusion reactions to occur at a sufficient rate requires a temperature in the region of 200 million degrees, and to obtain net energy production, the fuel density multiplied by the confinement time has to be larger than about 10^{21} nuclei per cubic meter multiplied by seconds. In magnetic confinement, the time involved is of the order of seconds and the density of the plasma is in the range 10^{20} to 10^{21} nuclei per cubic meter—many times less dense than air. In inertial-confinement fusion, the time involved is a few tenths of a billionth of a second and the density of the plasma has to reach 10^{31} nuclei per cubic meter—many times more dense than lead. The term ignition is used in both magnetic- and inertial-confinement fusion, but with different

Fusion, Second Edition.

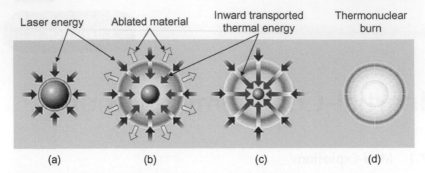

Laser energy　　Ablated material　　Inward transported thermal energy　　Thermonuclear burn

(a)　　　　(b)　　　　(c)　　　　(d)

FIGURE 7.1　The four stages of a fusion reaction in an inertial-confinement capsule: (a) Laser heating of the outer layer. (b) Ablation of the outer layer compresses the capsule. (c) The core reaches the density and temperature for ignition. (d) The fusion reaction spreads rapidly through the compressed fuel.

meanings. Magnetic confinement aims at reaching steady-state conditions, and ignition occurs when the alpha particle heating is sufficient to maintain the whole plasma at a steady temperature. Inertial-confinement fusion is inherently pulsed—the fuel capsule is compressed to a very high density but only a small fraction of the fuel needs to be heated initially and ignition occurs when the fuel capsule starts to burn outward from a central hot spot.

It is convenient to express the ignition criterion for inertial confinement in terms of the radius of the capsule multiplied by its density. Knowing the temperature of the plasma allows us to calculate the expansion velocity of the capsule and thus to convert the time in the conventional criterion into a radius (Box 7.1). The minimum value of the product of density and radius depends on the overall system efficiency of converting fusion energy output into effective capsule heating. In order to reach the required value with a reasonable amount of driver energy, it is necessary to compress the solid fuel to very high density (Box 7.2).

Usually, solids and liquids are considered to be incompressible. The idea that they can be compressed is outside our everyday experience, but it can be done if the pressure on the capsule is large enough. Experimental results show that it is possible to compress deuterium capsules to densities more than 1000 times their normal density.

The key to compressing a capsule is to heat its surface intensely so that it evaporates (or ablates) very rapidly (Figure 7.2). Hot gases leaving the surface act like those leaving a rocket engine, and they apply large forces to the capsule. The thrust developed for a brief instant in capsule compression is 100 times that of a space shuttle launcher! If the capsule surface is heated on just one side, the force of the evaporating gas will accelerate the capsule in the opposite direction, just like a rocket. However, if the capsule is heated uniformly from all sides, these forces can generate pressures about 100 million times greater than atmospheric pressure and thus compress the capsule.

Box 7.1 Conditions for Inertial Confinement

The conditions required for ICF are determined by the requirement that the fusion energy produced in one pulse has to be larger than the energy required to heat the fuel to the ignition temperature. The fusion energy (17.6 MeV per reaction) is extracted after each pulse and converted into electricity, and some of this energy has to be used to heat the fuel for the next pulse. For breakeven, with T in keV as in Box 4.3

$$\frac{1}{4}n^2\overline{\sigma v}\,17.6 \times 10^3 k\varepsilon\tau > 3nkT$$

$$n\tau > 6.82 \times 10^{-4}(T/\overline{\sigma v})\varepsilon^{-1}\,\mathrm{m}^{-3}\,\mathrm{s}$$

Note that τ is the pulse duration (the capsule burn time) and ε is the overall efficiency of converting fusion energy in the form of heat into effective capsule heating. This is the expression that John Lawson derived in the 1960s—and he was thinking in terms of pulsed magnetic-confinement fusion, so he assumed that all the fusion energy would be converted from heat into electricity, with $\varepsilon \approx 0.33$, and used for ohmic heating of the next pulse. At $T \approx 30$ keV, this gives the well-known expression $n\tau > 1 \times 10^{20}\,\mathrm{m}^{-3}\,\mathrm{s}$. However, ε is much smaller than 0.33 for ICF because of the need to convert heat into electricity and electricity into driver energy and then to couple the driver energy into effective fuel heating.

An ICF fuel capsule is compressed initially until it reaches a very high density (Box 7.2). Under optimal conditions, most of the compressed fuel stays at a relatively low temperature (100 to 200 eV) and only a small "hot spot" in the core reaches a much higher temperature. Then the core starts to burn and the burn propagates outward through the compressed capsule, which expands in radius r as it gets hot. The plasma density ($n \sim 1/r^3$) and fusion power ($P_F \sim n^2$) fall rapidly as the capsule expands, and we assume that the capsule stops burning when its radius has expanded by 25%, so $\tau \approx r/4v_i$. The expansion velocity of the capsule is determined by the ions (the electrons are held back by the heavier ions) and $v_i = 2 \times 10^5\,T^{0.5}\,\mathrm{m}^{-3}\,\mathrm{s}$ (for a 50:50 mixture of deuterium and tritium). Thus,

$$nr > 545(T^{1.5}/\overline{\sigma v})\,\varepsilon^{-1}\,\mathrm{m}^{-2}$$

The optimum is at $T \approx 20$ keV, where

$$T^{1.5}/\overline{\sigma v} \approx 2.1 \times 10^{23}\,\mathrm{keV}^{1.5}\,\mathrm{m}^{-3}\,\mathrm{s}$$

and

$$nr > 1.15 \times 10^{26}\,\varepsilon^{-1}\,\mathrm{m}^{-2}$$

This expression can be written in terms of the mass density $\rho = n \times 4.18 \times 10^{-27}\,\mathrm{kg\,m}^{-3}$. Thus

$$\rho r > 0.48\,\varepsilon^{-1}\mathrm{kg\,m}^{-2}(\text{or } 0.04\,\varepsilon^{-1}\mathrm{g\,cm}^{-2})$$

This is the condition for breakeven (when the fusion energy output is just sufficient to heat the next capsule). A fusion power plant has to produce a net output of power, and the condition is more stringent, as discussed in Box 13.7.

Box 7.2 Capsule Compression

If the capsule has radius r_0 before compression and is compressed by a factor c to $r = r_0/c$ before igniting, the compressed particle density will be $n = c^3 n_0$ and the mass density $\rho = c^3 \rho_0$. Using the results from Box 7.1, taking $\rho_0 \approx 300\,\text{kg m}^{-3}$ as the uncompressed density of a mixture of solid deuterium and tritium and expressing r_0 in millimeters and ε as a percentage, the condition for the minimum compression is

$$c^2 > 160/(r_0 \varepsilon)$$

For example, a capsule with uncompressed radius $r_0 = 1\,\text{mm}$ and conversion efficiency $\varepsilon = 1\%$ would require compression $c \approx 13$ to reach breakeven. Clearly, it is desirable to work at high conversion efficiency, but this is limited by the driver technology. The maximum capsule size is constrained by the maximum explosion that can be safely contained within the power plant. The fusion energy released by burning a capsule with uncompressed radius r_0 (in millimeters) is 4.25×10^8 r_0^3 joules, which is equal to $0.425\, r_0^3\,\text{GJ}$. So a DT fuel capsule with radius of $1\,\text{mm}$ contains fusion energy that is equivalent to about $10\,\text{kg}$ of high explosive.

Note that the calculations in this box and Box 7.1 contain significant simplifications. It is important that only the core of the compressed capsule is heated to ignition temperature. When the core ignites, the fusion reaction propagates rapidly through the rest of the compressed fuel. This reduces the energy invested in heating the capsule, but account also has to be taken of the energy needed for compression and of incomplete burn-up. Sophisticated numerical computer models are used to calculate the behavior of compressed capsules.

7.2 Using Lasers

Until the beginning of the 1960s there was no power source large enough to induce inertial-confinement fusion in a controlled way. The development of the laser suggested a method of compressing and heating fuel capsules on a sufficiently fast time scale. A laser is a very intense light source that can be focused down to a small spot size and turned on for just the right length of time to compress and heat the capsule. The principle of the laser (Box 7.3) was proposed by Arthur Schalow and Charles Townes at the Bell Telephone Laboratories in 1958, and the first working laser was built by Theodore Maimon in 1960. This was just at a time when magnetic-confinement fusion had reached something of an impasse, so laser enthusiasts saw an alternative approach to fusion. As early as 1963, Nicolai Basov (Figure 7.5) and Alexandr Prokhorov at the Lebedev Institute in Moscow put forward the idea of achieving nuclear fusion by laser irradiation of a small target. Some laser physicists thought that they could win the race with magnetic confinement to obtain energy *breakeven*, where the energy produced equals the initial heating energy.

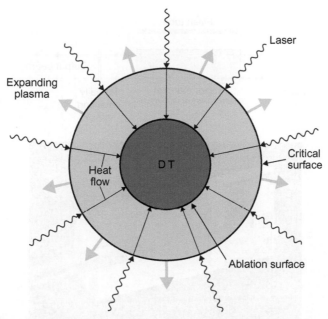

FIGURE 7.2 The ablation of a capsule by incident laser light creates a plasma expanding outward that applies a force in the opposite direction, compressing the capsule. Laser light cannot penetrate the dense plasma beyond a layer known as the *critical surface*. The laser energy is absorbed close to the critical surface and carried into the ablation surface by heat conduction.

Box 7.3 The Laser Principle

A laser (Figure 7.3) produces coherent light at a well-defined wavelength with all the photons in phase. To produce laser light, it is necessary to have a suitable medium in which atoms or molecules can be excited to a *metastable* level above the ground state. A system is said to be in a *metastable state* when it is poised to fall into a lower-energy state with only slight interaction. It is analogous to being at the bottom of a small valley when there is a deeper valley close by or to an object being on a shelf—where a small displacement can lead to its falling down to the floor.

The atoms are excited typically by a *flashlamp*, which when energized emits an intense burst of incoherent white light. Some photons from the flashlamp have the right wavelength to excite the atoms in the laser. Once a sufficient population of excited atoms in the metastable state has been accumulated, they can all be induced to decay at the same time by a process called *stimulated emission*. A single photon, of the same wavelength that is emitted, causes one excited atom to decay, producing a second photon and an avalanche process develops. The emitted photons have the same phase and propagation direction as the photon triggering the emission. Mirrors are placed at either end of the lasing medium so that each photon can be reflected back and forth, causing all the excited atoms to

(a)

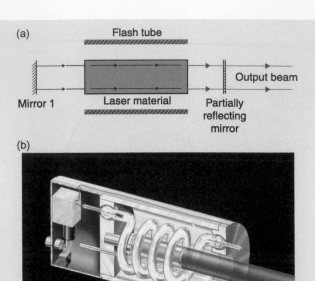

FIGURE 7.3 (a, b) Schematics of a simple laser system.

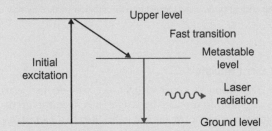

FIGURE 7.4 The energy levels of an atom in which lasing is produced by a two-stage process—excitation to the upper level, followed by prompt decay to a metastable level, sets up the conditions for stimulated emission.

decay in a time that is typically nanoseconds. The mirrors form an optical cavity, resonant with the desired wavelength of the laser, and the cavity can be tuned by adjusting the separation between the mirrors.

Suitable lasing media, where metastable excited states can be produced, include solids, liquids, gases, and plasmas. Normally, the excitation of the metastable state is a two-stage process, where the atom is initially excited to a higher-energy state that then decays to the metastable state (Figure 7.4). The laser light is coherent and can be focused to a very small spot, producing very high power densities. The first lasers tended to be in the red or infrared wavelengths, but suitable media have been found for lasers in the visible, ultraviolet, and even

X-ray wavelengths to be produced. It is also possible to double or triple the frequency of the laser light using special optical materials that have nonlinear properties. It is particularly important for ICF to bring the wavelength of the light from a neodymium laser ($1.053\,\mu m$) into a part of the spectrum ($0.351\,\mu m$) where the compression is more effective.

FIGURE 7.5 Nicolai Basov (1922–2001), along with Charles Townes and Alexandr Prokhorov, were awarded Nobel Laureate in Physics in 1964 for their development of the field of quantum electronics, which led to the discovery of the laser. Basov was the leader of the Soviet program on inertial-confinement fusion for many years.

The energy of the early lasers was too small, but laser development was proceeding rapidly, and it seemed likely that suitable lasers would soon be available. Basov and colleagues in Moscow reported the generation of thermonuclear neutrons from a plane target under laser irradiation in 1968, and a group in France demonstrated a definitive neutron yield in 1970. The expertise that had been developed in designing the hydrogen bomb included detailed theoretical models that could be run on computers to calculate the compression and heating of the fuel. This close overlap with nuclear weapons had the consequence that much of the inertial-confinement research remained a closely guarded secret, very reminiscent of the early days of magnetic confinement. Some details were released in 1972 when John Nuckolls (Figure 7.6) and his collaborators at the Lawrence Livermore National Laboratory in California outlined the principles in their landmark paper published in the international science journal *Nature*. Optimists predicted that inertial-confinement fusion might be achieved within a few years, but it soon became clear that it was not so easy.

Estimates of the amount of laser energy required to compress and heat a fusion capsule have increased several-fold since the early days, and present

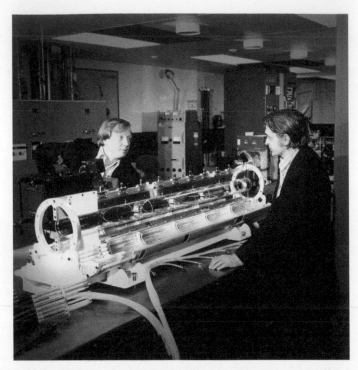

FIGURE 7.6 Physicists John Emmett (left) and John Nuckolls were the key Lawrence Livermore National Laboratory pioneers in laser and fusion science and technology. John Emmett co-invented the multi-pass laser architecture still in use today. John Nuckolls was the lead author of the landmark paper on inertial-confinement fusion in 1972 and a strong proponent of ICF in the US. He was director of the Lawrence Livermore National Laboratory from 1988 to 1994.

calculations show that at least 1 million joules will be required. This requires very advanced types of laser, and the most suitable ones presently available use a special glass containing the rare element neodymium. The laser and its associated equipment fill a building larger than an aircraft hanger. These lasers consist of many parallel beams that are fired simultaneously and focused uniformly onto the small fusion capsule. Typical of this type of laser was the NOVA facility at the Lawrence Livermore National Laboratory in the US (Figure 7.7).

NOVA came into operation in 1984 and had 10 beams. It was capable of producing up to 100,000 (10^5) joules in a burst of light lasting a billionth (10^{-9}) of a second (Figure 7.8). For that brief instant, its output was equivalent to 200 times the combined output of all the electricity-generating plants in the US. Large lasers have also been developed at the University of Rochester in the United States, at the Lebedev Institute in Russia, at Osaka University in Japan, at Limeil in France, and at other laboratories around the world. A new laser (at the National Ignition Facility, NIF) with state-of-the-art technology

FIGURE 7.7 The NOVA laser facility at the Lawrence Livermore National Laboratory. The central sphere is the target chamber, and the large pipes contain the optical systems bringing in the multiple laser beams.

FIGURE 7.8 A miniature star created inside the NOVA laser target chamber when an experimental capsule is compressed and heated by laser beams.

and energy more than 10 times that of existing facilities has been built and in 2010 started to operate at Livermore in the United States; a similar facility is nearing completion near Bordeaux in France. We discuss these latest developments in Chapter 12.

A neodymium laser produces an intense burst of light in the infrared part of the spectrum—just outside the range of visible light that can be seen by the human eye. However, this wavelength is not ideal for capsule compression. When the outer surface of a capsule is evaporated, it forms a high-density plasma that screens the interior of the capsule. Laser light in the infrared spectrum is reflected by this plasma screen, but light at a shorter wavelength, in the ultraviolet part of the spectrum, can penetrate deeper. In order to take advantage of the deeper penetration, the wavelength of the neodymium laser light is shifted from the infrared to the ultraviolet using specialized optical components, but some energy is lost in the process. Other types of laser, for example, lasers using krypton fluoride gas, are being developed to work directly in the optimum ultraviolet spectral region, but at present they are less powerful than neodymium lasers.

One of the problems inherent in compressing capsules in this way is that they tend to become distorted and fly apart before they reach the conditions for fusion. These distortions, known as instabilities, are a common feature of all fluids and plasmas. One particular form of instability that affects compressed capsules (known as the *Rayleigh-Taylor instability*) was first observed many years ago in ordinary fluids by the British physicist Lord Rayleigh. It occurs at the boundary between two fluids of different density and can be observed if a layer of water is carefully floated on top of a less dense fluid, such as oil. As soon as there is any small disturbance of the boundary between the water and the oil, the boundary becomes unstable and the two fluids exchange position so that the less dense oil floats on top of the denser water. A similar effect causes the compressed capsule to distort unless its surface is heated very uniformly. A uniformity of better than 1% is called for, requiring many separate laser beams, each of which has to be of very high optical quality, uniformly spaced all around the surface of the capsule.

An ingenious way of obtaining a high degree of uniformity was developed at Livermore around 1975, although the details remained secret until many years later. The capsule is supported inside a small metal cylinder that is typically a centimeter across and made of a heavy metal, such as gold, as in Figure 7.9. This cylinder is known as a *hohlraum*, the German word for "cavity." The laser beams are focused through holes onto the interior surfaces of this cavity rather than directly onto the capsule. The intense laser energy evaporates the inner surface of the cavity, producing a dense metal plasma. The laser energy is converted into X-rays, which bounce around inside the hohlraum, being absorbed and reemitted many times, rather like light in a room where the walls are completely covered by mirrors. The bouncing X-rays strike the capsule many times and from all directions, smoothing out any irregularities in

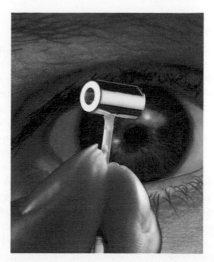

FIGURE 7.9 An experimental cavity, or hohlraum, of the type used for indirect-drive experiments. Intense laser light is shone into the cavity through the holes and interacts with the wall, creating intense X-rays. These X-rays heat the capsule, causing ablation, compression, and heating.

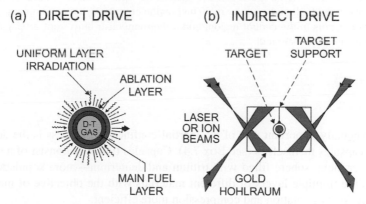

FIGURE 7.10 Comparison of the principles of direct and indirect drive. Uniform irradiation is produced by many laser beams in the direct-drive case and by the production of X-rays at the wall of the hohlraum in the indirect case. The targets are similar in size in the two cases, but the direct drive has been shown enlarged to illustrate the typical structure.

the original laser beams. Also, the X-rays can penetrate deeper into the plasma surrounding the heated capsule and couple their energy more effectively than longer-wavelength light. Some energy is lost in the conversion, but the more uniform heating compensates for this. This approach is known as *indirect drive*, in contrast to the *direct-drive* arrangement, where the laser beams are focused directly onto the capsule. Both approaches (Figure 7.10) are being studied in inertial-confinement experiments.

Box 7.4 Capsule Design

The ICF target capsule is generally a spherical shell filled with low-density DT gas ($<1.0\,mg\,cm^{-3}$), shown schematically in Figure 7.10a. The outer layer could be a plastic shell, which forms the ablator, and on the inside wall of the shell a layer of frozen deuterium and tritium (DT) forms the main fuel. The laser driver energy is deposited rapidly on the ablator layer, which heats up and evaporates. As the ablator evaporates outward, the rest of the shell is forced inward to conserve momentum. The capsule behaves as a spherical, ablation-driven rocket. The compression is achieved by applying a laser pulse that is carefully shaped in time so that it starts off at a low intensity and then increases to a maximum. A larger, thinner shell that encloses more volume can be accelerated to higher velocity than a thicker shell of the same mass, but the maximum ratio of capsule radius to shell thickness that can be tolerated is limited by Rayleigh-Taylor instabilities. The required peak implosion velocity determines the minimum energy and mass required for ignition of the fusion fuel. The smoothness and uniformity of the capsule surface are important factors in determining the maximum compression that can be reached. The typical surface roughness must be less than 100 nanometers ($10^{-7}\,m$).

   In its final, compressed state, the fuel reaches pressures up to 200 gigabars and ideally consists of two regions, a central hot spot containing between 2% and 5% of the fuel, and a cooler main fuel region containing the remaining mass. Ignition occurs in the central region, and a thermonuclear burn front propagates outward into the main fuel.

An equally important part of the inertial-confinement process is the design of the capsules themselves (see Box 7.4). Capsules typically consist of a small plastic or metal sphere filled with tritium and deuterium—more sophisticated targets use multiple layers of different materials with the objective of making the processes of ablation and compression more efficient.

It is important to avoid heating the capsule core too strongly during the early part of the compression phase because much more energy is then needed to compress the hot plasma. Ideally, a small spot in the core of the capsule should just reach the ignition temperature at the time of maximum compression. Once the core of the capsule starts to burn, the energy produced by the fusion reactions will heat up the rest of the capsule and the fusion reaction will spread outward. A more sophisticated scheme has been proposed by Max Tabak and colleagues at Livermore, where an ultrafast laser is fired to heat and ignite the core after it has been compressed. This technique, known as *fast ignition* (Box 7.5), may reduce the overall energy requirements for the driver. Another approach, known as *shock ignition*, has been recently developed and is discussed in Box 12.3.

Box 7.5 Fast Ignition

The conventional schemes (direct and indirect drive) for inertial confinement use the same laser for both compression and heating of the capsule. Fast ignition uses one laser to compress the plasma, followed by a very intense fast-pulsed laser to heat the core of the capsule after it is compressed. Separating the compression and heating into separate stages could lead to a reduction in the overall driver energy requirements.

Ultraviolet laser light cannot penetrate through the dense plasma that surrounds the compressed capsule core, and the key to fast ignition is a very intense, short-pulsed laser that might be able to bore its way in. Laser light focused onto a very small spot with intensity in the range 10^{18} to 10^{21} Wcm^{-2} produces a relativistic plasma with beams of very high-energy electrons moving at close to the speed of light—so fast that they can in fact keep up with the laser beam. The basic idea is that the relativistic electron beam would be able to punch through the compressed fuel and deliver sufficient energy into the core to heat it to ignition. Relativistic electron beams are produced when laser beams are focused onto target foils, but instabilities might break up these beams and prevent them from reaching the core of a compressed capsule. This is currently an active area of experimental and theoretical research.

An ingenious way around the problem has been suggested. A small cone is inserted into the fusion capsule to keep a corridor free of plasma during the compression, along which the fast-ignition laser pulse can propagate to the center of the compressed fuel. This idea was tested successfully in 2001 by a joint Japanese and UK team, using the GEKKO XII laser at Osaka University, Japan, to compress a capsule and a fast laser delivering 10^{14} W (100 terawatts) to heat it. Fast lasers at least 10 times more powerful—in the *petawatt* (10^{15} W) range—will be needed to approach breakeven and they are being developed in several laboratories (see Chapter 12).

7.3 Alternative Drivers

The big neodymium lasers presently at the forefront of ICF research are very inefficient—typically less than 1% of electrical energy is converted into ultraviolet light, and there is a further loss in generating X-rays in the hohlraum. A commercial ICF power plant will require much higher driver efficiency (see Box 13.7); otherwise, all its output will go into the driver. Using light-emitting diodes instead of flashlamps to pump a solid-state laser would improve the efficiency and permit the rapid firing rate needed for a power plant—but diode systems are extremely expensive at the moment, and costs would have to fall dramatically. Other types of lasers might also be developed, and ICF research is looking at alternative drivers using beams of energetic ions and intense bursts of X-rays.

Low-mass ions, such as lithium, are attractive as drivers because they can be produced efficiently with energies in the range needed to compress and heat

FIGURE 7.11   An open-shutter photo of the Z facility at Sandia National Laboratories in Albuquerque showing the "arcs and sparks" produced when the energy storage bank is fired.

a capsule. Systems to accelerate and focus intense beams of light ions have been developed at Sandia National Laboratories, but it has proved difficult to achieve adequate focusing, and the power intensities required for an ICF driver have not been obtained. Another difficulty is that the focusing elements have to be a very short distance from the target, while in order to survive the effects of the fusion mini-explosion, these elements should be at least several meters away.

Heavy-ion beams, such as xenon, cesium, and bismuth, are also being studied as drivers. Heavy ions need to be accelerated to a much higher energy than light ions, about 10,000 MeV, but the current would be smaller, and focusing to a small spot would be easier. However, the beam current requirements are still many orders of magnitude beyond what has been achieved in existing high-energy accelerators. The heavy-ion approach to ICF is certainly interesting and promising, but testing the basic physics and technology will require the construction of a very large accelerator.

The Sandia Z facility, which had been built to study light-ion beams, is now testing a different configuration that might drive ICF by producing an intense burst of X-rays from an ultrafast pinch discharge in an array of thin tungsten wires. Electrical energy is stored in a large bank of capacitors and is released through a special circuit that delivers a current of 20 million amps in a sub-microsecond pulse (Figure 7.11). The wires vaporize, forming a metal plasma that is accelerated by the enormous electromagnetic forces to

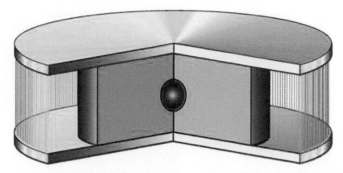

FIGURE 7.12 Schematic of the dynamic hohlraum concept. The ICF capsule is surrounded by low-density plastic foam at the center of the array of thin tungsten wires. An ultrafast pinch discharge drives a metal plasma inward at very high velocity, vaporizing the foam and forming its own hohlraum around the capsule.

implosion velocities as high as 750 kilometers per second ($7.5 \times 10^5 \, \text{ms}^{-1}$). In one arrangement, known as a *dynamic hohlraum*, the tungsten wire array surrounds a high-gain ICF capsule (Figure 7.12) in a block of low-density plastic foam. The imploding wire array forms a metal plasma shell that acts as the hohlraum wall, trapping the X-ray radiation that is formed when the plasma interacts with the foam and ablates the outer shell of the capsule.

7.4 The Future Program

A demonstration of ignition and burn propagation is the next important goal of the inertial-confinement program. The National Ignition Facility (NIF) at the Lawrence Livermore National Laboratory in California has already started operating, and the Laser Megajoule (LMJ) facility near Bordeaux, France, is nearing completion. We discuss these ongoing developments in Chapter 12.

Chapter 8

False Trails

8.1   Fusion in a Test Tube?

The world was taken by surprise in March 1989 when Stanley Pons and Martin Fleischmann held a press conference at the University of Utah in Salt Lake City to announce that they had achieved fusion at room temperature "in a test tube." Although Pons and Fleischmann were not experts in fusion, they were well known in their own field of electrochemistry, and Martin Fleischmann was a Fellow of the prestigious Royal Society. Their claims quickly aroused the attention of world media, and "cold fusion" made all the headlines. It seemed that these two university scientists with very modest funding and very simple equipment had achieved the goal sought for decades by large groups with complex experiments and budgets of billions of dollars. The implications and potential for profit were enormous—if fusion could be made to work so easily and on such a small scale, fusion power plants soon would be available in every home to provide unlimited energy.

Within days, the results were being discussed animatedly in every fusion laboratory around the world. Many fusion scientists suspected that something must be wrong—the details were sparse, but from the little that was known, these results seemed to violate fundamental laws of physics. Such skepticism is the norm in scientific circles whenever unexpected results are published, but in this case, when it came from established experts in fusion, it was regarded as the malign jealousy of opposing vested interests.

Pons and Fleischmann had used the process of *electrolysis* (Box 8.1), which is well known and can be demonstrated easily in any high school laboratory. An electrolysis cell, shown schematically in Figure 8.1, is a vessel containing a liquid (for example, water) and two metal electrodes. Passing an electric current between the electrodes ionizes the water, releasing hydrogen at one electrode and oxygen at the other. The cells used by Pons and Fleischmann contained *heavy water*, where deuterium took the place of normal hydrogen, and the electrodes were of a special metal—*palladium*. It had been known for more than a century that hydrogen is absorbed in metals like palladium and can reach very high concentrations, with one or more atoms of hydrogen for every atom of metal. Pons and Fleischmann's objective was to use electrolysis to pack deuterium into the palladium to such high concentrations that atoms might get close enough for fusion.

Fusion, Second Edition.

Box 8.1 Electrolysis

Electrolysis is the passing of an electric current through a liquid that contains electrically charged ions, known as an *electrolyte*. Water, being covalent, has very few ions in it, but the addition of a dilute acid or a salt, such as sodium chloride, provides sufficient ions to pass significant current. If sulfuric acid is used, the ions produced are H^+ and $(SO_4)^-$. The positive hydrogen ions are attracted to the negative cathode, where they pick up an electron, normally combine to form a molecule, and are released as a free gas. The negative sulfate ions go to the positive anode, where they are neutralized, but then react with water to produce more sulfuric acid and release oxygen. The sulfuric acid acts as a catalyst, and the result is the splitting up of water into hydrogen and oxygen. It was known that if an electrode such as titanium, tantalum, or palladium, which reacts exothermically with hydrogen, is used as the cathode, the metal becomes charged with hydrogen. Instead of being released as free gas, the hydrogen enters the metal and diffuses into the lattice. Hydrides can form with concentrations of up to two hydrogen atoms per metal atom.

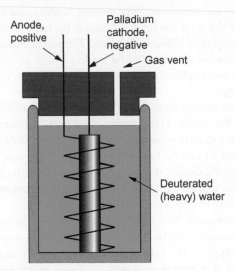

FIGURE 8.1 A schematic diagram of the electrolysis cells used by Pons and Fleischmann. Deuterated water (also known as heavy water) is contained in a glass vessel into which two metal electrodes are inserted. A current is passed through the water, and the deuterium is ionized and accumulates at the palladium cathode. It was claimed that energy was released from the cathode due to the occurrence of fusion reactions in it.

In fact, the idea was not so new—something similar had already been tried in Germany in the 1920s and again in Sweden in the 1930s, but these earlier claims for cold fusion had been abandoned long ago. Surprisingly, a second group, led by Steven Jones and Paul Palmer, was working on a similar topic in parallel to Pons and Fleischmann, only 30 miles away at Brigham Young University, but they were much more cautious in their claims. The competition between the two groups was one reason that prompted the hurried press conference.

The electrolysis cells got hot, and the temperature increase was measured in order to compare the energy that came out with the energy put into the cell by the electric current. The measurements were delicate and great care was needed to account for everything, and there were many complicating factors. Pons and Fleischmann claimed that more energy was released than was put in—a very startling result, and they ascribed the additional source of energy to nuclear fusion. It was natural that there should be skepticism about this interpretation. The process of hydrogen absorption in metals had been studied for many years, and it was known that the absorbed atoms are much too far apart for fusion ever to take place. It would take much more energy than normally can be found in a metal to force the deuterium nuclei close enough together for fusion. What could provide such energy? One explanation argued that the high concentration of deuterium in the palladium might cause cracking of the metal, a known effect, and that this could produce a large electric field, which might accelerate the deuterons to high-enough energies to cause them to fuse.

It was difficult for other scientists to obtain enough information to scrutinize the claims. The results had been presented at a press conference, allowing just the barest of details to be revealed, rather than by following the recognized procedure of publishing the results in a scientific journal, which would demand more serious evidence. A detailed description of the equipment was not released because the authors intended to take out a patent on the process, which naturally was thought to be very valuable. This reluctance to give details of the experiments and to respond openly to questions inevitably aroused suspicions.

Later, when Pons and Fleischmann submitted a paper describing their results to the prestigious international journal *Nature*, referees asked for more details, but the authors then withdrew the paper on the grounds that they did not have time to provide the requested information. One concern that the referees expressed was that proper control experiments had not been carried out. An obvious control would have been to compare identical cells, one with heavy water and one with ordinary water. The probability of fusion in normal hydrogen is negligible, and there should have been a clear difference between the cells if fusion really was taking place, as claimed, in deuterium. Although such a control experiment was so obvious, Pons and Fleischmann never presented any evidence to show that they had done one.

There were other discrepancies and grounds for concern. A genuine deuterium fusion reaction would produce other distinctive products that could be measured—tritium, helium, and neutrons—as well as energy. Pons and Fleischmann did not discuss the helium and tritium products at all, and initially their evidence for neutrons relied indirectly on measurements of energetic gamma rays. However, Richard Petrasso's group at the Massachusetts

Institute of Technology pointed out that the gamma-ray measurements presented at the press conference had the wrong energy—a peak was shown at 2.5 MeV, whereas the gamma ray from deuterium fusion should be at 2.2 MeV. This small but important discrepancy in the original data was never explained, but in later presentations the peak in the spectrum was shown at the correct energy. Later, when measurements of the neutron flux were shown, the number of neutrons turned out to be very much smaller than expected and was not consistent with the amount of fusion energy that was claimed. In fact, if the neutron radiation had been commensurate with the claimed heat output, it would have killed both Fleischmann and Pons! It was difficult to measure very small neutron fluxes against the background radiation from cosmic rays and naturally occurring radioactive elements. Some experts in neutron measurements thought that there might be problems with the instruments that had been used.

However, it is not in the nature of the scientific community to ignore a result just because no explanation is immediately apparent, especially when the implications are so immense. Groups large and small the world over set out to replicate the sensational results. The equipment was relatively simple and inexpensive, and many laboratories had the necessary expertise to repeat the experiments. Although experimenters were hampered by the lack of published details, intelligent guesses could be made as to what the equipment looked like, and many people started to try to see if they could get evidence of cold fusion. The measurements turned out to be more difficult than first thought. The amount of energy produced was really small and subtle variations in the design of the cell could upset it. Indeed, Pons and Fleischmann admitted, when pressed, that even in their own laboratory the results were not reproducible. In some cells it took a long time, often tens or hundreds of hours, before the energy anomaly was observed. This might have been explained by the time required to build up to the required density of deuterium in the metal, but in some cells nothing was observed at all. This kind of irreproducibility made it difficult to convince the world at large that there was a genuine effect. Under some conditions it could be attributed to an uncontrolled variable; possibly, the results were sensitive to an unknown parameter that was not being maintained constant.

With the difficulty of the experiments and the irreproducibility of the results, some groups inevitably began to get positive results and others found negative results. An interesting sociological phenomenon occurred. Those with "positive" results thought that they had succeeded in replicating the original results and looked no further—they rushed off to publish their results. But those who got "negative" results naturally thought that they might have made a mistake or had not got the conditions quite right. They were reluctant to publish their results until they had repeated and checked the measurements. This had the effect that apparently positive results were published first—giving the impression that all who attempted the experiment had confirmed the Pons and Fleischmann results. This type of effect had occurred previously with controversial scientific data. Nobel Prize winner Irving Langmuir, an early pioneer in the area of plasma physics, had studied the effect and classified it as one of the hallmarks of what he called "pathological science."

The negative results and the mountain of evidence against cold fusion were building up steadily. In spite of a large degree of skepticism that fusion was possible under these conditions, the implications were so important that many major laboratories had decided to devote substantial resources to repeating and checking the experiments. The Atomic Energy Laboratory at Harwell in the UK made one of the most careful studies. Martin Fleischmann had been a consultant at Harwell, and he was invited to describe his results in person and to give advice on how to obtain the right conditions for energy release. But even after months of careful research, scientists at Harwell could find no evidence for cold fusion. All other mainstream laboratories reached similar conclusions. The US Energy Research Advisory Board set up a panel to investigate the claims, but no satisfactory scientific verification was forthcoming. In some cases, high-energy outputs were recorded, but the necessary criteria for the occurrence of fusion were never properly fulfilled.

More than 20 years later, there are still a few passionate devotees who continue to believe in cold fusion, but convincing evidence is lacking. One must conclude that this was a sad episode in science. Two reputable scientists allowed themselves to be pressured into presenting their results prematurely by the desire to get patents filed and to claim the credit for making a momentous discovery.

8.2 Bubble Fusion

In 2002, the idea of fusion in a beaker was given another brief fillip when Rusi Taleyarkhan and colleagues at the Oak Ridge National Laboratory in the US claimed to have produced nuclear fusion in *sonoluminescence* experiments. They created very small bubbles (10 to 100 nm in diameter) in deuterated acetone (C_3D_6O) using 14 MeV neutrons and then applied sound waves that forced the bubbles first to expand to millimeter size and then to collapse. During the collapse, the vapor in the bubble heats up and produces a flash of light. It had also been speculated that nuclear fusion might be induced.

Taleyarkhan and colleagues claimed to have detected both 2.45 MeV neutrons and production of tritium, characteristic of the DD reactions, which were not observed when normal (nondeuterated) acetone was used. The result was so startling that the Oak Ridge management asked two independent researchers at the laboratory, Dan Shapira and Michael Saltmarsh, to try and reproduce the result, but they were unable to find any evidence for either the neutron emission or tritium production, even though they used a more efficient detector than Taleyarkhan had used. Later work by Yuri Didenko and Kenneth Suslick at the University of Illinois, who were studying how the acoustic energy is shared, showed that endothermic chemical reactions during the sonic compression would make it exceedingly difficult to reach the temperatures required for nuclear fusion to take place. Bubble fusion appears to be another false trail.

8.3   Fusion with Mesons

A completely novel approach to the fusion of light elements had occurred to Charles Frank at Bristol University, UK, in 1947. A group at Bristol, led by Cecil Powell, had been studying cosmic rays using balloons to carry photographic plates high up into the stratosphere. Cosmic rays entering the Earth's atmosphere from outer space were recorded as tracks on the plates, leading to the discovery of many new atomic particles, including what is now called the *mu-meson*, or *muon*. The muon is 207 times heavier than an electron. Although discovered originally in cosmic rays, muons can now be produced in particle accelerators.

A muon behaves rather like a very heavy electron, and Charles Frank realized that it would be possible to substitute a muon for the electron in a hydrogen or deuterium atom. According to the rules of quantum mechanics, the muon would have an orbit around the nucleus with a radius inversely proportional to its mass. The muonic atom, as it is now called, would be only 1/207 the size of a normal hydrogen atom. This means that such a muonic deuterium atom could approach very close to another deuterium or tritium atom and have the possibility of fusing and releasing energy. An equivalent way of looking at the situation is to say that the negative charge of the muon, in close orbit around the nucleus, shields the positive charge, allowing two nuclei to get close enough to each other to fuse (Figure 8.2).

There are two problems to any practical application of this idea. The first is that muons are unstable, with a lifetime of only 2.2 microseconds. The second is that it takes more energy to produce a single muon in an accelerator than the energy released by a single DT fusion reaction—about 1000 times more in fact. However, all was not yet lost. Frank realized that there was the possibility that the muon would be released after the DT fusion reaction, since the alpha particle would go off with high velocity, freeing the muon to initiate another fusion reaction. The muon would act like a catalyst, and the process is known as *muon-catalyzed fusion*. Obviously, the key to making this viable as a source of energy is that each muon must be able to catalyze at least 1000 fusion events in its short lifetime in order to generate more energy than has to be invested in making it.

In the Soviet Union, Andrei Sakharov took up Frank's idea. But his calculations of the rate of muon-catalyzed fusion at first appeared to indicate that the process was too slow to enable a net gain in energy within the lifetime of the muon. Meanwhile, the first observation of muonic fusion came in 1956 as a by-product of an accelerator experiment by Luis Alverez in Berkeley, California. Dedicated muonic-fusion experiments were subsequently conducted in the Soviet Union, first with DD and later with DT, demonstrating that the catalytic

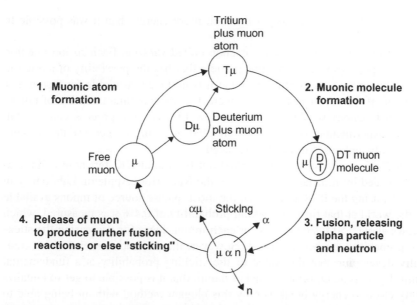

FIGURE 8.2 A schematic of the four stages of muon-catalyzed fusion. (1) The negative muon is captured by positive deuterium or tritium nuclei to form a muonic atom. The process tends to form muonic tritium atoms because they are slightly more stable than deuterium atoms. (2) The muonic tritium atom then combines with an ordinary deuterium molecule to form a complex muonic molecule. (3) The small size of the muon orbit partially screens the positive charges on the nuclei, allowing them to get close enough to each other to undergo a fusion reaction, forming an alpha particle and a neutron. (4) The alpha particle normally flies off, leaving the muon to repeat the process, but in some cases it sticks to the muon, preventing further reactions.

process works in principle. At around the same time, two Soviet physicists, Sergei Gerstein and Leonid Ponomarev, realized that the original theory was inadequate and that a muonic DT molecule could form much more rapidly than had previously been realized. There was thus the possibility that even in the short life of the muon it would be possible to have many fusion reactions.

This work led to renewed interest in the subject, and new experiments were undertaken to determine just how many fusion reactions per muon could be achieved. Experiments led by Steven Jones (who was later involved in the cold fusion work at Brigham Young University) at Los Alamos Laboratory in New Mexico used a gold-lined stainless steel vessel containing deuterium and tritium that could be controlled over a wide range of temperature and pressure. The theory indicated that the molecule formation would be faster at high pressure, and results published in 1985 showed that increasing the pressure of the DT mixture increased the number of fusions per muon to about 150. However, this was still far short of the number required for net production of energy. The parallel with "hot" fusion of the period, whether by magnetic confinement or by inertial confinement, was striking. It was possible to release fusion energy, but it was still necessary to put in much more energy than it was possible to get out.

The main problem is a phenomenon called *sticking*. Each fusion reaction creates a positively charged alpha particle that has the possibility of attracting and trapping the negative muon so that it is no longer available for fusion. For example, if the sticking probability were 1%, the maximum number of muon-catalyzed fusions would be 100, too low to make the process economical. Recent experimental work has concentrated on trying to measure the sticking probability and to see if it can be reduced.

In 2005, experiments were carried out by a joint UK–Japanese collaboration, headed by Kanetada Nagamine at the Rutherford Appleton Laboratory in the UK, using the ISIS accelerator, the most intense source of muons available in the world at that time. A sticking probability of 0.4% was measured, which allows about 250 fusion events for each muon. Ideas for decreasing the sticking probability are being studied. However, the realization seems to be gradually developing that this value of the sticking probability is a fundamental limit. It seems an unfortunate fact of nature that it is possible to get so tantalizingly close to energy breakeven by this elegant method without being able to achieve it.

Chapter 9

Tokamaks

9.1 The Basics

Chapter 5 describes the basic principles of magnetic confinement, but the story stopped in 1968 at the international fusion conference in Novosibirsk (Figure 9.1), just at the time when tokamaks had overtaken toroidal pinches and stellarators. The only cause to doubt the tokamak results was that the temperature of the tokamak plasma had been measured indirectly. To put the issue beyond all doubt, a team of British scientists was invited to go to Moscow the following year to measure the temperature with a new technique based on lasers (see Box 9.4). When the results were published, even the most skeptical fusion researchers were convinced. The leading American fusion laboratory, at Princeton, moved quickly, converting its biggest stellarator into a tokamak.

FIGURE 9.1 A group of Soviet and British scientists during the Novosibirsk conference in 1968, at which it was agreed that a team from the UK would go to Moscow to confirm the high plasma temperature claimed for the tokamak, using the new technique of scattering of laser light. From left to right: Bas Pease, Yuri Lukianov, John Adams, Lev Artsimovich, Hugh Bodin, Mikhail Romanovski, and Nicol Peacock.

Fusion, Second Edition.

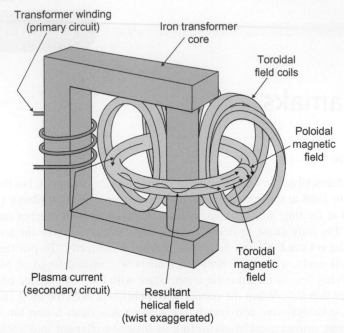

Transformer winding
(primary circuit)

Iron transformer
core

Toroidal
field coils

Poloidal
magnetic
field

Toroidal
magnetic
field

Plasma current
(secondary circuit)

Resultant
helical field
(twist exaggerated)

FIGURE 9.2   A schematic view of a tokamak showing how the current is induced in the plasma by a primary transformer winding. The toroidal magnetic field due to the external coils and the poloidal field due to the current flowing in the plasma combine to produce a helical magnetic field.

This was followed by the construction of many new tokamak experiments in the fusion laboratories of the US, Europe, Japan, and, of course, the Soviet Union. Thereafter, tokamaks set the pace for magnetic-confinement fusion research.

The basic features of a tokamak are shown in Figure 9.2. There are two main magnetic fields. One of them, known as the *toroidal field*, is produced by a copper coil wound in the shape of a torus. The second magnetic field, the *poloidal field*, is generated by an electric current that flows in the plasma. This current is induced by a transformer action similar to that described for the pinch system in Chapter 5. The two magnetic fields combine to create a composite magnetic field that twists around the torus in a gentle helix. Superficially, a tokamak resembles a toroidal pinch, which also has two similar fields. The main difference is in the relative strength of the two components. In a pinch, the poloidal field is much stronger than the toroidal field, so the helix is very tightly twisted, but in a tokamak it is the other way round—the toroidal field produced by the external coils is typically about 10 times stronger than the poloidal field due to the plasma current.

The main effect of the toroidal field is to stabilize the plasma. The most serious instabilities have to bend or stretch this magnetic field, which acts like the steel bars in reinforced concrete (although plasma is more like jelly than concrete). The fact that a tokamak has stronger "reinforcing bars" than a pinch has the effect of making it much more stable. Stellarators also have a strong toroidal field and share the inherent stability of the tokamak. However, because they have no plasma current, they require additional complex external coils to generate the poloidal magnetic field. These coils are expensive to build and must be aligned with great accuracy. So one of the reasons tokamaks overtook stellarators and stayed out in front is simply that they are easier and cheaper to build.

9.2   Instabilities

Trying to contain hot plasma within a magnetic field is like trying to balance a stick on one's finger or to position a ball on top of a hill—any small displacement grows with increasing speed. Plasma is inherently unstable and always tries to escape from the magnetic field. Some types of instability cause the sudden loss of the whole plasma; others persist and reduce the energy confinement time. Although these instabilities make it difficult to reach and maintain the conditions for fusion, it is important to stress that there is no danger, in the sense of causing explosions or damage to the environment. Plasma that escapes from the magnetic field cools quickly as soon as it touches a solid surface.

Even with the strong toroidal magnetic field, tokamak plasmas become unstable if either the plasma current or the plasma density is increased above a critical value. The plasma extinguishes itself in a sudden and dramatic manner, known as a *disruption* (Box 9.1). Disruptions can be avoided to some extent by careful selection of the operating conditions. The maximum current and density depend on the tokamak's physical dimensions and on the strength of the toroidal magnetic field (see Box 10.1). Disruptions occur when the magnetic field at the plasma edge is twisted too tightly. Increasing the plasma current increases the poloidal magnetic field, enhances the twist, and makes the plasma unstable. Increasing the toroidal field straightens out the twist and improves stability. The effect of plasma density is more complicated. In simple terms, increasing density causes the plasma edge to become cooler because more energy is lost by radiation, as discussed later. Because hot plasma is a much better electrical conductor than cool plasma, the cooler edge squeezes the current into the core, increasing the twist of the magnetic field by reducing the effective size of the plasma.

Some tokamak instabilities are subtler and do not completely destroy the plasma but result in loss of energy. For example, the temperature in the center of a tokamak usually follows a regular cycle of slow rises and rapid

Box 9.1 Disruptions

A major disruption in a tokamak is a dramatic event in which the plasma current abruptly terminates and confinement is lost. It is preceded by a well-defined sequence of events with four main phases. First, there are changes in the underlying tokamak conditions—usually an increase in the current or density, but not always clearly identified. When these changes reach some critical point, the second phase starts with the increase of magnetic fluctuations in the plasma, whose growth time is typically of the order of 10 ms. The sequence then passes a second critical point and enters the third phase, where events move on a much faster time scale—typically of the order of 1 ms. The confinement deteriorates and the central temperature collapses. The plasma current profile flattens, and the change in inductance gives rise to a characteristic negative voltage spike—typically 10 to 100 times larger than the normal resistive loop voltage. Finally, there is the current quench—the plasma current decays to zero at rates that can exceed 100 MA/s.

Disruptions cause large forces on the vacuum vessel, and these forces increase with machine size. In big tokamaks, forces of several hundred tons have been measured, and a power plant would have to withstand forces at least an order of magnitude higher. One of the factors determining when a disruption occurs is the amount by which the magnetic field lines are twisted. The twist is measured by the parameter q, known as the *safety factor*, which is the number of times that a magnetic field line passes the long way around the torus before it returns to its starting point in the poloidal plane—large q indicates a gentle twist and small q a tight twist. Usually, the plasma becomes unstable when the q is <3 at the plasma edge. (See Box 10.1 for a more detailed discussion.) Attempts to control disruptions have met with only limited success. Applying feedback control to the initial instability has delayed the onset of the third phase, but only marginally. Power-plant designers have proposed that disruptions should be avoided at all costs, but this seems unrealistic. A possible remedy is the rapid injection of impurities to cool the plasma by radiation before the onset of the current quench.

falls. A graph of temperature against time looks a bit like the edge of a saw blade and gives the name *sawteeth* to this instability (Box 9.2). Sawteeth are linked to the twist of the magnetic field in the center of the plasma. Very small-scale instabilities seem to appear all the time in tokamaks. Too small to be easily seen, they are held responsible for reducing plasma confinement (see Box 10.3). Tokamak instabilities are a very complex subject, and they are still far from being fully understood; they are similar in many ways to the instabilities in the weather, with which they have much in common, because both are controlled by the flow of fluids and show turbulent behavior.

Box 9.2 Sawteeth

The central temperature often varies in a periodic way, increasing in a linear ramp to a peak value, dropping abruptly, and then starting to increase again (Figure 9.3). This feature is described as *sawteeth*, for obvious reasons.

From observations at different radii, a flattening of the central temperature profile is observed when the temperature drops. There is also an increase in the temperature further out, indicating a sudden outflow of energy. A model proposed by Boris Kadomtsev from the Kurchatov Institute in Moscow was initially successful in explaining the behavior in terms of a *magnetic island*, which grows and tangles up the field lines. Eventually the tangled field sorts itself out by reconnecting—in effect, the field lines snap and rejoin in a more ordered way. The reconnection allows plasma in the hot central region to mix with cooler plasma. However, recent measurements of sawteeth reveal many details that are in disagreement with the Kadomtsev model, and, although attempts have been made to improve the theory, no model of sawteeth instability has yet been generally accepted.

It used to be assumed that sawteeth were triggered when the safety factor q reached unity in the center of the plasma (theory predicts that this is unstable). When techniques were developed to measure q directly inside a tokamak, it came as a surprise to find that sometimes q could be less than 1.

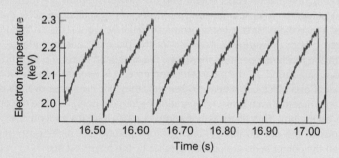

FIGURE 9.3 The variations of central temperature in a tokamak plasma, illustrating typical sawteeth.

9.3 Diagnosing the Plasma

The first stage of operation is getting the tokamak working (Box 9.3). The second stage is *diagnosing* the plasma—measuring its properties (Box 9.4). This is difficult because the plasma is so hot that it is not possible to put any solid probe deep into it. Much can be learned by analyzing the radiation emitted from the plasma: X-rays, light, radio emission, and neutral particles. More information is obtained by injecting laser beams, microwaves, and beams of

Box 9.3 Operating a Tokamak

The start-up phase of any new experiment is always exciting. There are many different components and systems, and all have to be working simultaneously before a plasma can be produced and heated. Although experience from the operation of other tokamaks is valuable, the operation of each one has its own individual characteristics. An analogy is that of learning to drive an automobile of the 1900s era. The driver knew roughly the right things to do, but it required trial and error to learn each car's idiosyncrasies, to learn how to change gear, and to learn how to make the car operate in the smoothest and most efficient manner. Similarly, tokamak operation has to be slowly optimized. It is known that there are limits to the plasma current, pressure, and density, above and below which the plasma will not be stable (for more detail, see Box 10.1). Where will these limits be in the new tokamak? Will the confinement time and the temperature be as predicted?

After the power supplies and control electronics have been checked, the first thing to do is to pump air out of the vacuum vessel and to clean it to reduce the impurities. Then the magnetic fields are turned on, deuterium gas is introduced, and a voltage is applied. This voltage causes the gas to break down electrically, become ionized, and form a plasma. As it becomes electrically conducting, a current begins to flow. The plasma current and the length of the current pulse are determined by the capacity of the power supplies. After each plasma pulse, the current falls, the plasma cools down, and the power supplies are recharged for the next pulse. The repetition rate for experiments is typically one pulse every 15 minutes. Once the plasma is being produced reliably, work starts on the detailed measurements of the various parameters of the plasma. Little by little, as confidence in the machine operation increases, the magnetic fields and the current in the plasma are raised. As the parameters increase, it is necessary to check all the mechanical and electrical systems to see that they are not being overstressed. It takes some months, even years, of operation before the operators find the optimal plasma conditions. This phase is common to many physics experiments and is analogous to the optimization of accelerator operation in high-energy physics. When all the components are operating reliably, the physics starts.

After typically 3 to 6 months of operation, the experiments stop so that maintenance can be carried out. Modifications and improvements are made in light of the previous operating experience. The shutdowns can last from a few weeks to over a year. Because the machines are operating near the limits of endurance of materials and because they are built as experiments rather than as production machines, improved operation has to be acquired gradually.

neutral atoms into the plasma and observing their attenuation or the light scattered or emitted from them. Many of these measurements involve sophisticated computer-controlled instruments and analysis. From the data obtained it is possible to calculate the density and the temperature of the electrons and ions, the energy confinement time, and many other properties. Measurements have

Box 9.4 Temperature Measurement

In fusion experiments it is important to be able to measure the properties of the plasma, including temperature, density, impurity concentration, and particle and energy confinement times. Because of the high temperatures involved, measurements have to be made remotely, usually by observing radiation from the plasma. Electron temperatures are measured by firing a laser beam into the plasma and detecting the light scattered by the electrons; a technique known as Thomson scattering. The scattered light is Doppler shifted in frequency due to the temperature of the electrons. By viewing different positions along the length of the laser path, a radial profile of temperature can be measured (Figure 9.4).

The temperature of the plasma electrons is determined from the spectrum of the scattered laser light. The local electron density can also be obtained from the total number of photons scattered from a specific volume. The number of photons scattered is generally small, so a high-energy laser (ruby or neodymium) pulsed for a short interval is employed. The laser may be pulsed repetitively to obtain the time evolution of the temperature. Electron temperature can also be measured from electron cyclotron radiation, emitted due to the acceleration of the electrons as they follow circular orbits in the magnetic field. Impurity ion temperatures can be measured directly using a high-resolution spectrometer to observe the Doppler broadening of a suitable spectral line. Because impurities have different charge states in different regions of the plasma, it is possible to measure temperatures in different radial positions. Hydrogen ions in the plasma core cannot be measured in this way because they are fully ionized and do not emit spectral lines. Normally, hydrogen ions and the impurities are at similar temperatures, and some correction can usually be made for any discrepancy between them.

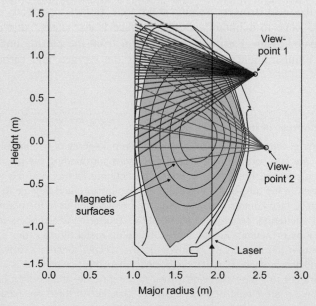

FIGURE 9.4 The Thomson scattering diagnostic for the DIII-D tokamak in San Diego. The laser beam passes through the plasma along a vertical line. Up to 40 detector channels receive light scattered from different positions in the plasma via the two viewpoints.

to be made at many different positions because of the variation in the plasma properties, from the hot and dense core to the relatively cool edge. Another type of investigation is studying the instabilities in the plasma to determine the conditions that induce the instabilities and the effect that the instabilities have on the loss of energy.

When the plasma has been properly diagnosed, the next stage is to compare its properties with the predictions of various theoretical models. In this way it is possible to get a better understanding of why the plasma behaves in the way it does. Only with this understanding is it possible to see how to improve the plasma and to get closer to the conditions required for a fusion power plant.

9.4   Impurities

Tokamak plasmas are normally contained in a stainless steel toroidal vacuum vessel that nests inside the toroidal magnetic coils. All the air inside the vessel is evacuated, and it is filled with low-pressure deuterium gas that is ionized to create plasma. Although ionized from pure deuterium gas, plasma can quickly become contaminated with other elements. These are known as *impurities*, and their main source is the interaction between the plasma and the material surfaces (Box 9.5).

Layers of oxygen and carbon atoms always cover surfaces in a vacuum, even under the most stringent vacuum conditions. In fact, a perfectly clean surface is impossible to maintain—a clean surface is so "sticky" that it quickly becomes coated again. Particles and radiation from the plasma bombard the wall. This bombardment dislodges some of the oxygen and carbon atoms

Box 9.5 Sources of Impurities

In a burning plasma, the fusion process is an internal source of helium ash. Other impurities are released from the material surfaces surrounding the plasma by a variety of processes. There are impurities, particularly carbon and oxygen, that are entrained in the bulk of the metal during manufacture and migrate to the surface. These surface contaminants are released by radiation from the plasma or as a result of sputtering, arcing, and evaporation. *Sputtering* is a process in which energetic ions or neutrals knock atoms, including metal atoms, from a surface by momentum transfer. *Arcing* is driven by the voltage difference between the plasma and the surface. *Evaporation* occurs when the power load is sufficient to heat surfaces to temperatures near their melting point—this is often localized in hot spots on prominent edges exposed to plasma. All three mechanisms are important at surfaces that are subject to direct plasma contact, such as the limiter or divertor, but generally the walls are shielded from charged particles and are subject only to sputtering by charge-exchange neutrals.

sticking to the surface and even knocks metal atoms out of the surface itself (Box 9.5). These impurity atoms enter the plasma, where they become ionized and trapped by the magnetic fields. Impurity ions like oxygen and carbon, with not too many electrons, radiate energy and cool the plasma most strongly near the plasma edge (Box 9.6). Too much edge cooling makes the plasma unstable, and it disrupts when the density is raised. In order to reach high plasma density, the concentrations of oxygen and carbon impurities have to be reduced to less than about 1%. Metal ions, especially those with many electrons, like tungsten, are responsible for much of the energy radiated from the core. Too much cooling in the core makes it difficult to reach ignition temperatures.

The effects of impurities are reduced by careful design and choice of materials for the internal surfaces of the tokamak. The internal surfaces are carefully cleaned; the toroidal vacuum chamber is heated up to several hundred degrees Celsius and conditioned for many hours with relatively low-temperature plasmas before a tokamak operates with high-temperature fusion plasmas. Soviet physicists had a neat term for this process; they referred to "training" their tokamaks. However, even with the best training, there were still impurities from the walls of the vacuum chamber in which the plasma was confined.

Box 9.6 Impurity Radiation

Impurities in tokamak plasmas introduce a variety of problems. The most immediate effect is the radiated power loss. A convenient parameter to characterize the impurity content is the *effective ion charge*, $Z_{eff} = \sum_i n_i Z_i^2 / n_e$, where the summation is taken over all the ionization states of all the ion species.

Line radiation from electronic transitions in partially ionized impurity ions is the most important cooling mechanism, but it ceases at the temperature where the atoms are fully ionized and lose all their electrons. Hence, radiative cooling is particularly significant in the start-up phase, when the plasma is cold, and also at the plasma edge. Low-Z impurities, such as carbon and oxygen, will lose all their electrons in the hot plasma core. But to reach ignition, one must overcome the radiation peak, which for carbon is around 10 eV. Next, the plasma must burn through the radiation barrier around 100 keV due to medium-Z impurities, such as iron and nickel, and that around 1 keV due to high-Z impurities, such as molybdenum and tungsten. A DT plasma with as little as 0.1% of tungsten would radiate so much power that it would be impossible to reach ignition.

From the point of view of radiation cooling, much larger concentrations of carbon, oxygen, and helium ash could be tolerated in a fusion power plant, but then the problem of fuel dilution arises. An impurity ion produces many electrons. In view of the operating limits on electron density and plasma pressure (see Box 10.1), this has the effect of displacing fuel ions. For example, at a given electron density, each fully ionized carbon ion replaces six fuel ions, so a 10% concentration of fully ionized carbon in the plasma core would reduce the fusion power to less than one-half of the value in a pure plasma.

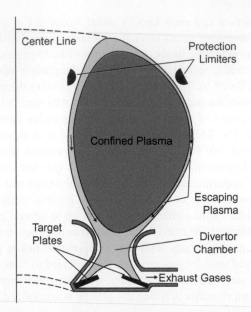

FIGURE 9.5   An example of a type of tokamak divertor at the bottom of the machine. The poloidal magnetic field generated by the plasma current is distorted by extra coils (not shown), which produce a magnetic field that is locally in the opposite direction to the poloidal field. This allows the plasma escaping from the confined region to stream down into the divertor chamber and to interact with a surface (the divertor target) that is remote from the confined hot plasma.

A very important feature of tokamaks is that there must be a well-defined point of contact between plasma and wall that can withstand high heat loads. This is called a *limiter* because it limits the size of the plasma. Limiters in the early tokamaks were usually made of metals, like tungsten and molybdenum, which can withstand high temperatures. However, because these metals cause serious problems if they get into the plasma, many tokamaks now use materials like carbon and beryllium, which have less impact as impurities. But carbon and beryllium bring other problems, and there is no ideal choice of wall material for a fusion power plant—this is one of the issues still to be resolved.

Suggestions had been put forward as early as the 1950s by Lyman Spitzer at Princeton of a method for mitigating this wall interaction in his stellarator designs. He proposed a magnetic-confinement system where the magnetic field lines at the edge of the plasma are deliberately *diverted* or led away, into a separate chamber, where they can interact with the wall. This system is called a *divertor*. Impurities produced in the divertor are inhibited from returning to the main chamber by restricting the size of the holes between the two chambers and because the flow of the main plasma from the main chamber tends to sweep impurities back into the divertor, rather like a swimmer in a strong current. When the stellarator fell out of favor, this idea was applied to the tokamak, with considerable success (Figures 9.5 and 9.6).

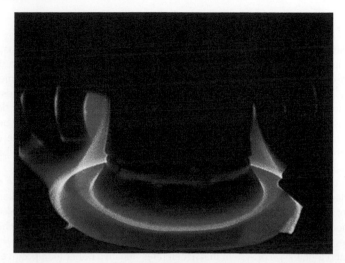

FIGURE 9.6 Plasma light emission from the divertor region in the ASDEX-U tokamak in Garching, Germany. The bright rings of light are the slots where the plasma makes the strongest contact within the director.

9.5　Heating the Plasma

The electric current in the plasma has two functions. In addition to generating the poloidal magnetic field, it heats the plasma, in the same way that an electric current flowing through a metal wire heats it up. This heating effect, known as *ohmic* or joule heating, was very efficient in early tokamaks and was one of the reasons for their success. However, as the plasma gets hotter, its electrical resistance falls and the current becomes less effective at heating. The maximum temperature that can be reached with ohmic heating is typically less than 50 million degrees—very hot by most standards, but still some way to go on the scale required for fusion.

The most obvious solution would appear to be to increase the plasma current. However, increasing it too much will cause the plasma to disrupt (Box 9.1), unless the toroidal magnetic field is increased proportionally (Box 10.1). The forces on the magnetic coils increase rapidly with field and set the upper limit to the strength of the toroidal field. Some tokamaks have pursued this approach to very high magnetic fields (see Section 10.7). A spin-off benefit is that the maximum plasma density and the energy confinement time increase with magnetic field. Early success gave rise to optimism that the conditions for ignition could be reached in this way—but there are severe engineering difficulties.

The need for additional plasma heating to give control of plasma temperature independent of the current has led to the development of two main heating techniques. The first technique, known as neutral-beam heating, uses powerful beams of energetic neutral deuterium atoms injected into the plasma (Box 9.7). The second plasma-heating technique uses radio waves (Box 9.8),

Box 9.7 Production of Neutral Heating Beams

The neutral beams used for plasma heating start life outside the tokamak as deuterium ions that are accelerated to high energies by passing them through a series of high-voltage grids (Figure 9.7). A beam of energetic ions cannot be injected directly into a tokamak because it would be deflected by the magnetic fields. The ion beam is therefore converted into neutral atoms by passing it through deuterium gas, where the ions pick up an electron from the gas molecules. The beams of neutral atoms are not deflected by magnetic fields and can penetrate deep inside the plasma, where they become reionized. After being ionized, the beams are trapped inside the magnetic fields and heat the plasma by collisions.

The energy required for the neutral beams has to match the size of the tokamak plasma and its density. The typical beam energy for present-day tokamaks is about 120 keV, but the neutral beams for the next generation of tokamaks and for fusion power plants require much higher energy, up to 1 MeV. At these higher energies, the process of neutralizing a positive ion beam becomes very inefficient, and a new type of system that accelerates negative rather than positive ions is required.

Much work has been done over many years in developing high-current sources in order to inject the maximum power into tokamaks. There is a theoretical limit to the amount of ion current that may be extracted through a single hole in the accelerating structure, and grids with multiple holes, up to more than 200, are used. One of the technical difficulties is removing the waste heat in this system. Currents of about 60 amps of ions are routinely generated at 120 keV, but the neutralization efficiency is only about 30%, leading to an injected power of just over 2 MW of neutrals per injector.

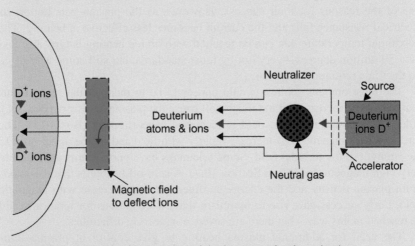

FIGURE 9.7   Schematic of a neutral beam injection system for plasma heating.

> **Box 9.8 Radiofrequency Heating**
>
> The highest radio frequency used for plasma heating is determined by the *electron cyclotron resonance*, $\omega \approx \omega_{ce}$, which depends only on the toroidal magnetic field. The resonance at 28 GHz/T (gigahertz per tesla) falls in the frequency range 60 to 120 GHz for most experiments, but frequencies up to 200 GHz will be required for a power plant. The free-space wavelength is in the millimeter waveband. Waves outside the plasma can be propagated in metal waveguides, and the launching antennas can be placed well back from the plasma edge. Electron cyclotron waves heat the electrons, which in turn transfer energy to the ions by collisions. The method may be used for both global and localized heating and has been applied to control sawteeth and disruptions. The application of electron cyclotron heating has been hindered by a lack of powerful, reliable high-frequency sources. But this is now being overcome, and these techniques will play an increasingly important role in fusion.
>
> Lower in frequency, the *ion cyclotron resonance*, $\omega \approx \omega_{ci}$, depends on the charge-to-mass ratio of the ion (Z/A) and on the toroidal magnetic field. The fundamental resonance lies at 15.2 (Z/A) MHz/T. The technology of power generation and transmission is well developed because the frequency range of interest for fusion, 40 to 70 MHz, is the same as that widely used for commercial radio broadcasting. In a tokamak, the toroidal field falls off as $1/R$, so the heating is localized at a particular value of *radius*, determined by the frequency. The physics of the heating process is quite complicated and requires either a plasma with two ion species (for example, a concentration of a few percent hydrogen minority in a deuterium plasma) or working at the second harmonic resonance $\omega \approx 2\omega_{ci}$. Antennas must be placed very close to the plasma edge for efficient coupling because the waves cannot propagate below a certain critical plasma density (typically about 2×10^{18} m^{-3}). Plasma interaction with these antennas, aggravated by the localized intense radiofrequency fields, can lead to a serious problem with impurity influx. In spite of its name, ion cyclotron heating is usually more effective in heating the plasma electrons than in heating the plasma ions. A combination of neutral beam heating and ion cyclotron heating allows some degree of control of both ion and electron temperatures.
>
> There is a third resonance frequency, midway between the ion and electron cyclotron frequencies, known as the *lower hybrid resonance*. It falls in the range 1 to 8 GHz, corresponding to free-space wavelengths in the microwave region of the spectrum. It has proved less effective for plasma heating but is used to drive currents in the plasma.

rather like a microwave oven, where food is heated by absorbing energy from a microwave source. The plasma version requires a bit more sophistication to select a radio frequency at which the plasma will absorb energy. The most important frequency ranges are linked to the natural resonance frequencies of ions and electrons in the toroidal magnetic field. The ion resonance frequency is in a range similar to that used for radio and television transmitters, and the electron resonance frequency is in a range close to that used for radar. Using a combination of neutral-beam and radio-wave heating provides a flexible arrangement and allows some control over the temperature of both the electrons and the ions in the plasma.

These heating techniques were introduced to tokamaks in the 1970s and soon plasma temperatures began to rise toward the range required for nuclear fusion. At first it seemed that the fusion goal was close to hand, but the temperature rise was less than expected. It was already known that energy loss from a tokamak plasma was faster than expected theoretically, but careful studies now showed that the loss increased further as the heating power was increased. In other words, the energy confinement time decreases as the heating is applied. An analogy is a house with central heating where the windows open more and more as the heating is turned up. The room can still be heated, but it takes more energy than if the windows stay closed. Plasma temperatures in the range required for fusion can be reached, but it takes much more heating than originally expected.

This was bad news, but there was a ray of hope when some experiments started to show better confinement than others. A new operating regime with high confinement was discovered in the ASDEX tokamak in Germany by combining a divertor with neutral-beam heating; it became known as the H-mode (Box 9.9).

Box 9.9   L- and H-Modes

During neutral beam heating of the ASDEX tokamak in 1982, Fritz Wagner and his colleagues found that, under certain conditions, there was an abrupt transition in the plasma characteristics and the energy and particle confinement improved. The behavior was unexpected, but subsequently this type of transition was observed in other tokamaks with divertors. It has also been observed, but with more difficulty, in tokamaks with limiters. The state of the plasma after the transition has been called the H-mode (for high confinement), to distinguish it from the usual L-mode (for low confinement). The improvement in energy confinement is typically a factor of 2.

The onset of the H-mode requires the heating power to be above a certain threshold. The transition is first observed at the edge of the plasma, where there is a rapid increase in both the edge density and temperature, resulting in steep edge gradients. It can be simply thought of in terms of an edge transport barrier. The edge and central temperatures and densities rise, but the new equilibrium is not easily controlled. Usually, the edge region becomes unstable (Figure 9.8), with periodic bursts of plasma loss known as *edge-localized modes (ELMs)*.

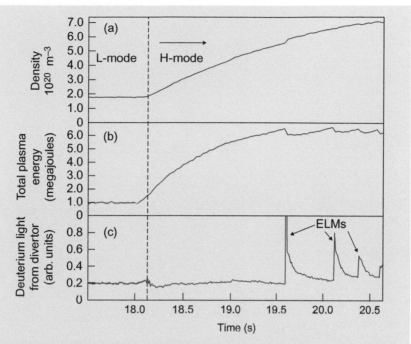

FIGURE 9.8   Illustration of typical behavior of the transition from L- to H-mode showing:
(a) the density, (b) the plasma energy, and (c) the light emitted by deuterium atoms in the
divertor, with the spikes characteristic of ELMs.

Chapter 10

From T3 to ITER

10.1   The Big Tokamaks

The success of the newly built tokamaks in the early 1970s encouraged fusion scientists to develop ambitious forward-looking plans. In the wake of the oil crisis of 1973, alternative energy sources enjoyed renewed popular and political support and the prospects for funding fusion research improved. However, extrapolating to the plasma conditions for a fusion power plant from the relatively small tokamaks then in operation was a very uncertain procedure. Plasma temperatures had reached less than one-tenth of the value required for ignition, and energy confinement times had even further to go. It was clear that powerful heating would be required to reach the temperature range for ignition, but heating techniques (see Chapter 9) were still being developed, and it was too early to say how energy confinement would scale.

Although theoretical understanding of the energy and particle loss processes was lacking, experimental data from the early tokamaks showed that both plasma temperature and energy confinement time appeared to increase with plasma current. Though there was much scatter in the data, the projections available at that time suggested that a tokamak with a current of about 3 million amps (3 MA) might reach the point known as *breakeven*, where the fusion energy output matches the required heating input. This would still fall short of *ignition* (the condition where heating by alpha particles is sufficient to sustain the plasma without any external heating) but would be a crucial step toward making the task of designing a fusion power plant more certain.

The French TFR and Russian T4 tokamaks were the most powerful in operation in the early 1970s, with maximum plasma currents of about 400,000 amps. The Americans were building a 1 MA tokamak, known as the *Princeton Large Torus* (*PLT*), due to start operating in 1975. The Europeans decided on a bold step and in 1973 started to design a 3 MA tokamak—the *Joint European Torus* (*JET*). Too big and too expensive for a single fusion laboratory, JET became a collaborative venture, with most of the funding provided by the European Community. The design progressed rapidly, but political decisions came slowly. Both Britain and Germany wanted to host JET. The ministers most directly concerned were Anthony Wedgewood Benn and Hans Matthöfer, and neither would give way. The crisis came in October 1977 as the design team was beginning to disintegrate, and the resolution came

Fusion, Second Edition.
107

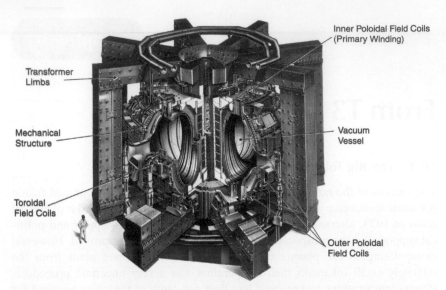

Inner Poloidal Field Coils
(Primary Winding)

Transformer
Limbs

Mechanical
Structure

Vacuum
Vessel

Toroidal
Field Coils

Outer Poloidal
Field Coils

FIGURE 10.1 A cutaway model of the JET tokamak, showing the iron transformer, the magnetic field coils, and the vacuum vessel in which the plasma is formed. Built as the flagship experiment of the European Community fusion program, JET remains the largest tokamak in the world.

about in a rather surprising way. On October 17 there was a spectacular rescue at Mogadishu, Somalia, of a Lufthansa plane that had been hijacked by the Baader-Meinhof gang. The UK had supplied special grenades used to stun the terrorists. In the general atmosphere of goodwill following the rescue, Chancellor Helmut Schmidt and Prime Minister James Callaghan appeared to come to an understanding. The decision to build JET at Culham, just south of Oxford in the UK, was finally made at a meeting of the European Community research ministers on October 25.

The initial European mandate to build a 3 MA tokamak was extended to 4.8 MA during the design phase. In fact, JET (Figure 10.1) had been well designed by the team, led by Paul Rebut (Figure 10.2), and was able to reach 7 MA. The contrast with earlier fusion experiments, many of which had failed to reach their design goals, was striking. Water-cooled copper coils provided JET's toroidal magnetic field. The coils were "D-shaped" in order to minimize the effects of electromechanical forces. Compared to a circular cross-section, the elongated shape allowed plasma with bigger volume and increased current, and later it allowed the installation of a divertor (Figure 10.3).

Peak electrical power for JET's magnets, heating, and other systems requires 900 MW, a substantial fraction of the output of a big power plant. Taking such a large surge of electrical power directly from the national electricity grid would have side-effects for other customers, so two large flywheel generators boost the power for JET. These generators, with rotors weighing 775 tons, are run up to speed using a small motor between tokamak pulses to store energy in the

FIGURE 10.2   Paul-Henri Rebut and Hans-Otto Wüster in the JET control room for the first plasma in 1983. Rebut (on the left) was the leader of the JET design team from 1973 to 1978, Deputy Director of JET from 1978 to 1987, Director from 1987 to 1992, and Director of ITER from 1992 to 1994. Wüster (on the right) was Director of JET from 1978 to 1987.

FIGURE 10.3   A wide-angle view (taken in 1996) inside JET. Most of the inner surfaces are covered with carbon tiles. The divertor can be seen as the annular slots at the bottom. The man wearing protective clothing is performing maintenance within the vessel.

rotating mass. When the rotor windings are energized, the rotational energy is converted to electrical energy. Delivering up to 2600 MJ of energy with a peak power of 400 MW, each generator slows down to half speed in a few seconds; after that, the power supplies connected directly to the grid take over.

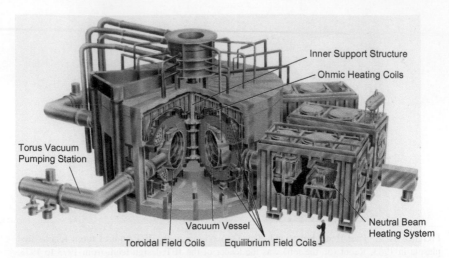

FIGURE 10.4   A cutaway model of the TFTR tokamak, built at Princeton, showing the magnetic field coils, the vacuum vessel, and the neutral beam heating system. It was the largest tokamak in the US fusion program until it closed down in 1998.

Europe's decision to design a big tokamak provoked a quick response from the Americans. Though slightly later in starting the design of their *Tokamak Fusion Test Reactor* (*TFTR*), they were quicker to reach a decision and to select a site, at Princeton University. With a circular plasma and slightly smaller dimensions but a higher magnetic field than JET, TFTR (Figure 10.4) was designed for a plasma current of 2.5 MA. The Soviet Union announced tentative plans to build a large tokamak, designated T-20, but was beginning to fall behind in the tokamak stakes due to priorities in other areas of science, and they later settled for a smaller experiment, T-15. Japan decided to build a big tokamak, known as JT-60, specified to operate at 2.7 MA. It was the last of the big tokamaks to start operation—in April 1985. After major modifications a few years later, JT-60U (as it was then renamed) came to match JET in performance. The dimensions of the three large tokamaks—JET, TFTR, and JT-60U—are compared in Figure 10.5. Keen competition among the three research teams provided a powerful stimulus, but there was always close collaboration and exchange of ideas and information.

10.2   Pushing to Peak Performance

The intense efforts to build TFTR came to fruition in the afternoon of Christmas Eve 1982, when, for the first time, a puff of hydrogen was injected into the torus and the magnetic coils energized to produce the first plasma. JET began operations about 6 months later, in June 1983. JET and TFTR quickly showed that they could operate reliably with currents of several million amps—much higher than previous tokamaks (Box 10.1). The gamble in

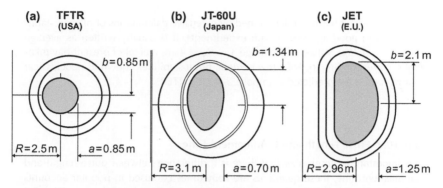

FIGURE 10.5 Comparison of the three large tokamaks built in the 1970s to demonstrate that fusion power increases with machine size and to get as close as possible to the conditions for net power production. (a) TFTR at Princeton, (b) JT-60U at Naka, Japan, and (c) JET at Culham, UK.

Box 10.1 Operating Limits

There are limits on the maximum plasma current, density, and beta (the ratio of plasma pressure to magnetic pressure) in a tokamak. Disruptions (see Box 9.1) usually determine the current and density limits; other instabilities set the beta limit.

The *current limit* is conservatively taken as equivalent to $q_a > 3$, where q_a is the value of the parameter q (known as the *safety factor*) at the plasma edge. q_a depends on the plasma minor radius a, major radius R, vertical elongation κ, toroidal magnetic field B, as well as the plasma current I. Approximately, q_a is proportional to $(aB/I)(a/R)\kappa$, and it can be seen that maximum current requires a torus that is vertically elongated (large κ) and "fat" (large a/R). Most tokamak designs push these geometric factors as far as possible within engineering and physics constraints—typical values are $\kappa \approx 1.8$ and $a/R \approx 1/3$ for a conventional tokamak and $a/R \approx 1$ for a spherical tokamak. Then the ratio I/aB is approximately a constant (typically $I/aB \approx 1.4$ in units of mega-amps, meters, and teslas for JET and ITER with $q_a \approx 3$). The maximum current I is proportional to aB and increases with both plasma minor radius and toroidal field (but there is a limit to the maximum toroidal magnetic field at about 6T in a large superconducting tokamak).

The *density limit* is determined empirically by taking data from a wide range of experiments. The commonly used *Greenwald limit* gives a maximum for the line-average electron density (this is close to, but not exactly the same as, the average density) $n_G = I/\pi a^2$, where density is in units of $10^{20}\,\mathrm{m}^{-3}$ and I is in MA. This can be rewritten in the form $n_G = (B/\pi a)(I/aB)$ to bring out the dependence on magnetic field ($n_G \sim B$) and plasma radius ($n_G \sim 1/a$), with I/aB approximately constant, as discussed above. It is important to note that this is an upper limit on the electron density and that the fuel ion density is reduced by the presence of ionized impurities that contribute electrons (see Box 9.6).

The *beta limit* has been determined by extensive calculations of plasma stability and is in good agreement with experiments. It is usually written as average β (in %) $= \beta_N(I/a)$. Values in the range $3 < \beta_N < 4$ are typical of present-day tokamaks, corresponding to β of a few percent, but instabilities known as neoclassical tearing modes are predicted to impose $\beta_N \approx 2$ in ITER.

Box 10.2 Pulse Length and Confinement Time

It is important to emphasize that there is a difference between pulse length and confinement time; these quantities are sometimes confused in popular accounts of fusion.

Pulse length is the overall duration of the plasma and usually is determined by technical limits on the magnetic field, plasma current, and plasma heating systems. In present-day experiments, the pulse length usually lasts several tens of seconds, but some experiments with superconducting magnets already have much longer pulse length. ITER, which is discussed in Chapter 11, will have a pulse length of several hundred seconds, and a fusion power plant will have a pulse length of many hours or days. In a tokamak with superconducting magnetic field coils, the pulse length is ultimately determined by the fact that some of the plasma current is driven inductively. A superconducting stellarator does not need a current in the plasma and so can be steady state.

Confinement time is a measure of the average time that *particles* (ions and electrons) or *energy* spends in the plasma, and this is generally much shorter than the pulse length. The *particle confinement time* is usually longer than the *energy confinement time* because energy is lost from the plasma by thermal conduction as well as by thermal convection. The energy confinement time is an important quantity that enters into the ignition condition (Box 4.3) and is defined as $\tau_E = E/P$. E is the total kinetic (i.e., thermal) energy of the plasma ($E \approx 3nkTV$, where n and T are, respectively, the mean density and temperature and V is the plasma volume), P is the total power into (or out of) the plasma, and k is Boltzmann's constant. When the plasma is in steady-state thermal equilibrium, the power in (from external heating and alpha particles) equals the power out. ITER will have $\tau_E \approx 3.7$ s and this is a modest extrapolation from present-day experiments (Box 10.4).

taking such a big step had paid off. The pulse lengths in the new machines were much longer than had been achieved before; in JET the pulses usually lasted 10 to 20 seconds, but on some occasions they were increased to over 1 minute (Box 10.2). Energy confinement time increased so much compared to previous experiments that it was now quoted in units of seconds rather than milliseconds. At last, confinement time had entered the range needed for ignition; this was a major milestone in fusion research.

For the first few years, JET and TFTR used only the ohmic-heating effect of the plasma current. This yielded temperatures higher than predicted from previous, smaller experiments, ranging up to 50 million degrees, but still too low for significant fusion power. Additional heating systems, which had been revised several times in light of results from smaller tokamaks, were added in stages and upgraded progressively over a period of years. Ultimately, each of the big tokamaks had over 50 MW of heating capacity. In JET, the heating was split roughly equally between neutral beams and radiofrequency heating, whereas TFTR and JT-60 used a higher proportion of neutral beam heating. Plasma temperatures reached the target of 200 million degrees; indeed, TFTR got as high as 400 million. However, as in smaller tokamaks, the confinement time got worse as the heating was increased.

There were some rays of hope that confinement times could be improved. The ASDEX tokamak in Germany had found conditions known as the *H-mode* (Box 9.9), with a dramatic improvement in the energy confinement when neutral beam heating was combined with a divertor. JET had been designed without a divertor, but fortunately the D-shaped vessel allowed one to be installed. The first divertor was improvised in 1986 using the existing magnetic coils. The results were impressive and produced a record value of the $nT\tau_E$ product (density $\times$ temperature $\times$ energy confinement time, Box 4.3), twice as good as the best previously obtained. Thus encouraged, the JET team set about designing and installing a purpose-built divertor. The circular shape of TFTR made it difficult to follow the same route, but the Princeton scientists found they could get impressive results if they carefully conditioned the torus and used intense neutral beam heating. The following few years proved to be extremely successful, with both machines giving encouraging scientific results. There was intense competition to produce increasingly higher values of the $nT\tau_E$ product, and the best values came within a factor of 5 of the target needed for ignition.

JT-60 was the only one of the three big tokamaks to have been designed at the outset with a divertor, but the first version took up a lot of space and seriously restricted the maximum plasma current. JT-60 was shut down in 1989 for installation of a new divertor. Taking advantage of the already large dimensions of the toroidal field coils, the divertor substantially changed the appearance of JT-60. It allowed an increase in the plasma current, up to 6 MA. The new version (renamed JT-60U) came into operation in March 1991 and began to play a major role in advanced tokamak research. As the JET and TFTR researchers turned their attention to operation with tritium plasmas, it was left to the JT-60U scientists to take the lead in pushing the performance in deuterium.

10.3 Tritium Operation

So far, fusion experiments had avoided using tritium and had operated almost exclusively in deuterium plasmas. This minimized the build-up of radioactivity in the tokamak structure due to neutron bombardment and thus made it easier to carry out maintenance and upgrades. However, it was important to gain experience with real operation in tritium plasmas, and both TFTR and

JET were designed with this in mind. Both tokamaks were housed inside thick concrete walls to shield personnel and had provisions for carrying out future maintenance work using tools that could be operated remotely. Encouraging results in deuterium plasmas pushed these plans forward.

In November 1991, JET became the first experiment to use a deuterium-tritium mixture. A large proportion of the JET team was in the control room for this exciting experiment. The first tests to check that the measurement systems were working properly used a concentration of only 1% tritium in deuterium. When the tritium concentration was raised to 10%, the peak fusion power rose to slightly more than 1 million watts (1 MW). At last, here was a clear demonstration of controlled nuclear fusion power in a significant quantity—enough to heat hundreds of homes, although only for a second or so. The results were close to those predicted from earlier deuterium experiments, and it was calculated that about 5 MW would be produced with the optimum concentration of 50% tritium. However, it was decided to postpone this step for a few more years in order to make it easier to install the new divertor.

The TFTR team was not far behind. In November 1993 they introduced tritium into their tokamak and quickly raised the concentration to the optimum 50%. In an intensive campaign between 1993 and 1997, TFTR raised the fusion output to more than 10 MW. When JET returned to tritium experiments in 1997, improved plasmas obtained with the new divertor allowed the fusion output to reach a record value of about 16 MW, lasting for a few seconds. This was still not a self-sustaining fusion reaction because it had more than 20 MW of external heating. When various transient effects are taken into account, the calculated ratio of the nuclear fusion power to the effective input power is close to 1—the long-sought conditions of breakeven. Figure 10.6 shows a summary of the JET and TFTR tritium results.

In addition to the direct demonstration of fusion power, the tritium experiments in TFTR and JET gave important scientific results. Plasma heating by the high-energy alpha particles from the DT reaction was measured. In JET, the alpha heating was about 3 MW, compared with 20 MW of external heating applied with the neutral beams. Although not yet enough to be self-sustaining, this was an encouraging demonstration of the principle of alpha particle heating. One fear that had been allayed was that some unforeseen effect might cause loss of the energetic alpha particles faster than they can heat the plasma. A welcome bonus was that energy confinement in tritium is slightly better than in deuterium. TFTR and JET also proved that tritium can be handled safely. JET tested the first large-scale plant for the supply and processing of tritium to a tokamak in a closed cycle, and the unburned tritium was reused several times in the tokamak.

10.4   Scaling to a Power Plant

The three big tokamaks, supported by the many smaller experiments, pushed magnetic confinement to conditions that are very close to those required for fusion. The results have steadily improved over time, from the early Russian

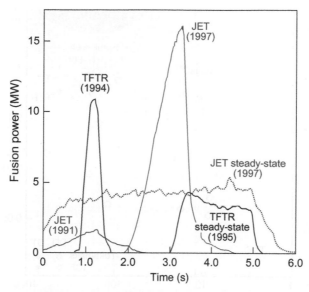

FIGURE 10.6 Summary of the fusion power produced with deuterium-tritium fuel in TFTR and JET during the period 1991–97. Plasmas with peak power lasting 1–2 seconds are compared with lower-power plasmas lasting for a longer time.

experiment T3 in 1968 to the present day. The $nT\tau_E$ product (Figure 10.7) has increased by more than three orders of magnitude and is within a factor of 5 of the value required for ignition. In separate experiments, temperature and density have reached or exceeded the values required for ignition. For operational convenience, most of these results are obtained in DD plasma and scaled to DT, but JET and TFTR have gone further and have values using DT plasma.

Fusion scientists extrapolate to the energy confinement times and other key parameters for future experiments on the basis of measurements made in many smaller experiments. It has proved very difficult to calculate the underlying particle diffusion and heat conduction from basic plasma theory. The basic processes that cause energy loss from a tokamak are known, as explained in Box 10.3, but the physics is so complex that actual loss rates cannot be calculated with sufficient accuracy to replace empirical data. In some ways, the problem is similar to that of evaluating the strength of a piece of steel—this cannot be calculated from first principles with sufficient accuracy to be of practical use. When engineers need to know the strength of steel in order to build a bridge, they actually measure the strength of small sample pieces and scale up to a larger size using their knowledge of how the strength depends on the length and cross-section of the samples. Similarly, plasma *scaling* studies establish the key parameters and quantify how confinement time depends on them (Box 10.4). The empirical scaling of energy confinement time from present-day experiments to a power plant is shown in Figure 10.8.

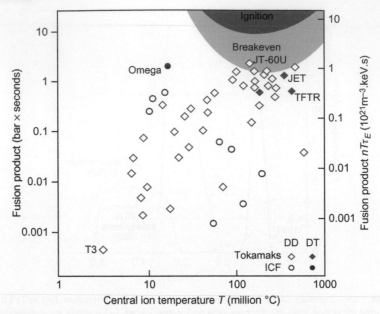

FIGURE 10.7 Summary of the progress toward the goal of controlled fusion since the Russian T3 experiment in 1968. The ion temperature T is plotted along the horizontal axis, and the $nT\tau_E$ fusion product is plotted along the vertical axes. JET, TFTR, and JT-60U have come very close to ignition conditions (top right-hand corner). The open symbols indicate data extrapolated to DT from measurements in DD; the solid symbols are actual measurements in DT. Inertial confinement data are included, although a direct comparison of the fusion product between inertial and magnetic confinement in this way has some complications.

Box 10.3 Understanding Confinement

The simple classical theory of diffusion across a magnetic field with cylindrical symmetry (Box 5.1) gave a diffusion coefficient of the form ρ^2/t_c, where t_c is the characteristic time between collisions and ρ is the Larmor radius in the longitudinal magnetic field. The extension to toroidal systems is known as *neoclassical theory* and takes into account the complex orbits of charged particles in a torus. In essence, the relevant step length is now determined by the poloidal magnetic field. Neoclassical theory has been developed in considerable detail and is thought to be correct—but it gives values of the diffusion coefficient that are generally smaller than those seen in experiments. Attempts to find refinements to the theory that remove the discrepancy have not met with success. It is now accepted that neoclassical theory describes a lower limit to diffusion that usually is overshadowed by higher losses due to some form of fine-scale turbulent fluctuations.

In spite of intensive efforts by all of the world's fusion programs, it has proved extremely difficult to model these fluctuations theoretically. This is a very complex problem, with different regimes of turbulence responsible for different aspects of plasma transport (particle diffusion, heat conduction, and so on) in different regions of the plasma—but these turbulent regimes interact with and influence each other. Thus a theory of a single regime of turbulence (and there are many such theories) is unable to provide a realistic model of the whole plasma. There has been encouraging progress in the past few years in developing a closer understanding of these processes and, in particular, in learning how to influence and control the turbulence in order to improve confinement. An example is the control of the distributions of plasma current and radial electric field in order to produce and optimize the so-called internal transport barrier.

In order to address the problem theoretically, it is necessary to solve a set of highly nonlinear equations (known as the *gyrokinetic* equations) that describe plasma in non-uniform magnetic fields. It is particularly difficult to solve these equations in regimes of saturated turbulence (a problem common with all other branches of physics that deal with highly nonlinear systems). The qualitative aspects of turbulence can be modeled, but quantitative estimates of diffusion rates are much more difficult.

An alternative computational approach follows the trajectories of all the individual particles in the plasma, taking into account their equations of motion and their mutual electromagnetic interactions. This is called the *particle-in-cell* method. But the long-range nature of the electromagnetic force means that each particle interacts with virtually all the other particles in the plasma—so it is necessary to solve a very large number of coupled equations for each time step. The most powerful computers presently available are unable to carry out the calculations needed to model these highly turbulent regimes and to derive diffusion coefficients with sufficient accuracy.

Box 10.4 Empirical Scaling

In the absence of an adequate theoretical understanding of confinement (Box 10.3), it is necessary to rely on empirical methods to predict the performance of future experiments. Data from many tokamaks over a wide range of different conditions are analyzed using statistical methods to determine the dependence of quantities like energy confinement time on parameters that include the current, heating power, magnetic field, and plasma dimensions.

In 1984, Robert Goldston proposed one of the best-known scaling models for tokamaks heated by neutral beams, $\tau_E \sim 1/P^{0.5}$, where P is the total heating power applied to the plasma. This convenient approximate value of the exponent has proved to be remarkably resilient over the intervening decades for a much wider range of plasma dimensions than in the original study and with other forms of heating.

The ITER scaling (Figure 10.8) is based on an extensive database of confinement studies in the world's leading tokamaks. In the form in which it is usually written, it predicts strong dependence of confinement time on inverse power ($\tau_E \sim 1/P^{0.65}$) and on plasma current ($\tau_E \sim I^{0.9}$) as well as on plasma size, magnetic field, and other parameters. However, a formulation in terms of the heating power and plasma current can be misleading because these parameters are also dependent on other parameters in the scaling expression. For example, the plasma current, minor radius, and magnetic field are linked by $I/aB \approx$ constant (see Box 10.1) and P can be written in terms of τ_E, density, and temperature (see Box 10.2). A revised formulation brings out the implicit dependence on plasma temperature ($\tau_E \sim 1/T^{1.25}$) that brings the optimum temperature for DT ignition in a tokamak down to around 10 keV (see Box 4.3).

The scaling for confinement time can be combined with the empirical limit for the maximum density (see Box 10.1) into an expression for the fusion triple product that shows that $nT\tau_E \sim a^2 B^3$ when quantities like I/aB and a/R are held constant. Thus it is clear that increasing the plasma radius and working at the highest possible magnetic field are the key steps in reaching ignition. There are, of course, many other parameters that need to be considered in optimizing the design of a fusion power plant, but these factors lie beyond the scope of this simplified account.

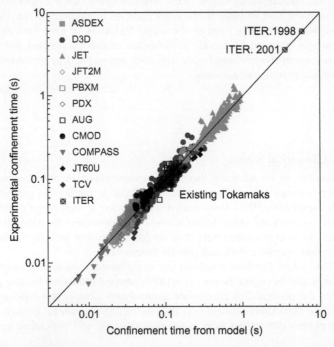

FIGURE 10.8 Data from a wide range of tokamaks in the international fusion program, showing the excellent agreement between experimental energy confinement time and the empirical scaling. Extrapolating this scaling, the energy confinement time is predicted to be 3.7 s for the present ITER design (2001) and 6 s for the earlier design (1998).

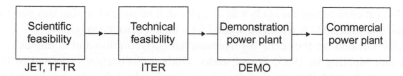

FIGURE 10.9   Outline of the development of the fusion energy program.

There is considerable confidence in this empirical approach to confinement-time scaling, which has been refined over a long period and is based on data from a very wide range of experiments.

An important parameter turns out to be the plasma current, and the extrapolation on the basis of this scaling for the confinement time indicates that a minimum of 20 MA will be needed for ignition in a tokamak. There are limits on the maximum current (see Box 10.1) for a given plasma radius and toroidal magnetic field, and a current of 20 MA requires a tokamak that is approximately three times bigger than JET.

10.5   The Next Step

To get to a commercial fusion power plant based on the tokamak concept will require three major steps (Figure 10.9). The first step—to show that fusion is *scientifically feasible*—has been achieved by the three big tokamaks: TFTR, JET, and JT-60U. The second stage—usually known as the *Next Step*—requires an even bigger experiment that will test most of the technical features needed for a power plant (including superconducting magnetic field coils and tritium production, as discussed in Chapter 13) in order to show that fusion is *technically feasible*. The third step, tentatively known as DEMO (which is also discussed in Chapter 13), will be to build a prototype fusion power plant producing electricity routinely and reliably to show that fusion is *commercially feasible*. Optimistic plans in the 1970s aimed at starting the Next Step in the 1980s and having the DEMO stage operating early in the 21st century, but various factors have combined to delay these plans by at least two decades.

Back in the late 1970s, with the construction of the big tokamaks already under way, the next task was to think about the design of the Next Step experiment. Groups in the United States, the Soviet Union, the European Community, and Japan started to design their own versions. They came together to collaborate on a study that was known as *INTOR*, the *International Tokamak Reactor*. The INTOR study did valuable work in identifying many of the technical issues for a fusion power plant and stimulated research programs to solve them. However, INTOR fell short on the physics specification—not surprisingly, because the three big tokamaks were still only at the construction stage and it was a very large extrapolation from existing smaller tokamaks. Estimates from the small tokamaks led INTOR to be designed around a plasma current of

8 MA, but even before the study was completed, it became clear that this was too small—at least 20 MA would be needed for a tokamak to reach ignition.

INTOR had set the scene for international collaboration and at the Geneva Summit Meeting in 1985, Soviet leader Mikhail Gorbachov proposed to US President Ronald Reagan that the "Next Step" tokamak experiment should be designed and built as an international collaboration. The United States responded positively and, in collaboration with Japan and the European Community, agreed to embark on the design phase of ITER. (The name *ITER,* pronounced "eater," has a dual significance—it is an acronym for *International Thermonuclear Experimental Reactor* and a Latin word meaning *the way.*) The initiative was received enthusiastically by the international tokamak community and conceptual design studies for ITER started in 1988. The tentative planning envisaged that construction of ITER would start during the 1990s—but there have been many frustrating delays in this ambitious schedule. The good news is that construction of ITER is now in progress. We discuss this important project in more detail in Chapter 11.

10.6   Continuing Research

Due to the long delay in the ITER program, JET and JT-60U have continued to be operated for much longer than originally expected, and in the meantime several new tokamaks have started operating. These experiments give the opportunity for trying a variety of new ideas that would, if successful, subsequently be tested at a larger scale on ITER. Many of the existing tokamaks have been used to model and study ITER operation. A series of experiments carried out in JET, DIII-D, ASDEX Upgrade, JT-60U, and other smaller tokamaks have attempted, as far as possible, to reproduce the conditions of an ITER discharge. They have resulted in improvements in the predictive capability of transport physics codes and have given confidence that burning plasmas in ITER will be achieved when operated at full toroidal field and full plasma current. The ramp-up to full current and the ramp-down following the burning phase have been shown to be critical, and techniques for controlling these phases have been developed. Another major area of technical development has been the building of divertor tokamaks with superconducting coils. The first large superconducting tokamaks were the *Tore Supra* machine at Cadarache in France and *T-15* in Moscow, but these have circular plasma cross-sections without divertors. *EAST* is a new large superconducting tokamak with a divertor that was built at the Institute for Plasma Physics in Hefei, China, in 2006, and *KSTAR* is a similar project at Daejeon in South Korea. EAST has produced plasmas with currents up to 0.8 MA and has demonstrated heating with radiofrequency waves. Lower current discharges have been extended up to 60 seconds. An upgrade of JT-60U to replace the existing water-cooled copper magnetic field coils with superconducting coils (Figure 10.10) is planned as part of the Broader Approach (see Box 13.6).

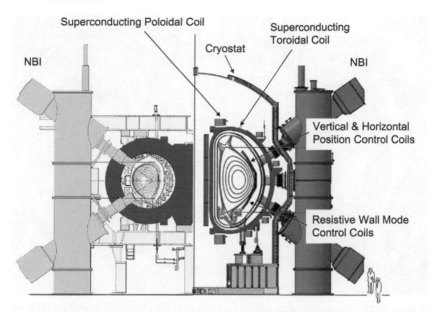

FIGURE 10.10 A cross-section through the JT-60U tokamak showing (on the left) the present machine (JT-60U) with water-cooled copper magnetic field coils and (on the right) the proposed upgrade (JT-60SA) with superconducting field coils.

One of the principal areas of study in recent years has been the investigation of new materials for the walls and divertors. The discovery that large quantities of hydrogen isotopes are retained in carbon components in the tokamak and the difficulty of removing them led to the realization that carbon is an unsatisfactory material from the point of view of tritium inventory. The relatively high erosion rate of carbon, principally due to chemical sputtering, is also a problem leading to material erosion from one component and deposition elsewhere within the tokamak. To try to find a solution to these problems, the walls and divertor target in the ASDEX Upgrade tokamak have been changed in stages from carbon to tungsten. This is a reversion to the materials used for limiters in early days of tokamak operation. The earlier materials had been abandoned when it was found that large amounts of tungsten or molybdenum in the plasma led to unacceptable amounts of radiation, which cooled the plasma excessively. However, operating at higher densities with cooler plasma edges showed that the tungsten contamination of the plasma core could be controlled. In a parallel development, JET has been extensively modified by installation of beryllium walls in the main tokamak chamber and tungsten targets in the divertor, in order to test the combination of wall materials that will be used in ITER. The installation of the *ITER-like Wall* in JET (Figure 10.11) was completed in 2011 and is showing stable plasmas and good impurity control. Beryllium and tungsten components are expected to retain a much lower amount of tritium than carbon.

FIGURE 10.11   The interior of the JET torus in mid-2011 after the installation of the ITER-like Wall. The walls are now protected with beryllium tiles and tungsten is used in the divertor. The robotic manipulator arm can be seen on the left.

10.7   Variations on the Tokamak Theme

Not all fusion scientists have been in agreement about the strategy of the route to fusion energy based on large tokamaks. Some have looked for quicker and cheaper ways to reach ignition by proposing smaller experiments to separate the fusion physics from the fusion technology. One approach used tokamaks that operate at very high magnetic fields, allowing higher plasma current to be squeezed into a plasma with small physical size. These are typified by the Alcator series of experiments at the Massachusetts Institute of Technology (MIT), the current version being Alcator C-Mod. The operating limits on a tokamak—namely plasma current, density, and pressure (beta)—scale very favorably with the toroidal magnetic field (Box 10.1). Increasing the toroidal magnetic field to more than 10 T allows higher plasma currents in a torus with smaller physical dimensions. The energy confinement time τ_E and the fusion triple product $nT\tau_E$ also scale very favorably with the toroidal magnetic field B (Box 10.4). However one problem with this approach is that the magnetic fields are too high for superconducting magnets, so copper coils have to be used and this technology does not lead directly to that of a power plant. Still, the proponents have argued that if ignition could be reached this way, it would allow much of the relevant physics to be explored and would give a tremendous boost in confidence that fusion would work. Several designs for high-field ignition experiments have been proposed along these lines, but only one has been built. It started operation in 1991 at the Triniti laboratory in Troitsk near Moscow but never reached its full performance, due to a combination of technical problems and reduced funding for scientific research after the break-up of the Soviet Union.

Proposals for high-field ignition experiments in the US and Europe have been discussed for many years but have failed to find support. A high-field tokamak called *Ignitor* was proposed in the 1970s by Bruno Coppi at MIT as a development of the *Alcator* series of experiments. The proposed design of Ignitor has a toroidal magnetic field of about 13 T and a plasma current of about 11 MA. Ignitor's aim is to explore the physics of burning fusion plasmas, as well as the conditions needed to achieve a controlled, self-sustaining reaction. The design has been examined and re-examined over many years without getting full funding, although technical development of components has been carried out in Italy. In 2010, an agreement was finally reached between Italy and Russia, with the approval of Prime Ministers Silvio Berlusconi and Vladimir Putin, to fund the design. It was reported in 2011 that Ignitor will be built at the Triniti laboratory in Troitsk as part of a major program to revitalize Russian scientific research.

Research on smaller tokamaks and other confinement geometries is still an important part of the fusion program to develop possible long-term alternatives. One line of research is the so-called *spherical tokamak*, where the tokamak concept is shrunk into a much more elongated and fatter torus, with a ratio of major to minor radius $R/a \approx 1$. This allows the spherical tokamak to operate at lower magnetic fields and higher plasma pressure relative to the magnetic pressure (Box 10.1). The geometry is achieved by dispensing with the usual arrangements of toroidal field coils in favor of a single current-carrying rod down the central vertical axis. This concept has been tested in several small experiments, and two large experiments of this type have been built, MAST at Culham and NSTX at Princeton, which can operate with plasma currents up to 1 MA. There are some doubts about the prospects for scaling-up the spherical tokamak concept to a power plant. The restricted space on the inside of the torus does not allow sufficient shielding to use superconducting magnetic field coils. Even copper coils and their insulation will be badly affected by neutron damage. However, there are proposals to build a compact neutron source based around a spherical tokamak in order to test components for future fusion power plants.

10.8 Stellarators Revisited

The most serious competition to the tokamak is the stellarator (Box 10.5), whose early development is discussed in Chapter 5. The stellarator configuration has the attraction of being able to confine plasma without the need for a current in the plasma, so it can be steady state. For a long time, stellarators lagged behind tokamaks in terms of their plasma performance for a variety of reasons. Stellarators are more difficult to build and so for many years they tended to be smaller and less powerful than the tokamaks of the same era. Until the techniques of powerful neutral beam and radiofrequency heating

Box 10.5 Stellarators

The early development of the stellarator is discussed in Chapter 5. As explained in Box 5.2, plasma cannot be confined in a purely toroidal magnetic field—the magnetic field lines have to be twisted as they pass around the torus. The proper name for the twist is a *rotational transform,* which is described in more detail in Box 10.6. In a tokamak, the poloidal component of the magnetic field that gives the twist is produced by the current flowing in the plasma, whereas in the stellarator the twist is produced by external coils. The stellarator configuration has the attraction of being able to confine plasma without the need for a current in the plasma and so it can be steady state.

The *classical stellarator* (Figure 5.7) has two independent sets of magnetic field coils. First there is a set of toroidal field coils, as in a tokamak, that produces the toroidal component of the magnetic field. The second set of coils is wound helically around the plasma and it sits inside the toroidal field coils. The helical coils are arranged in pairs, with currents flowing in opposite directions in adjacent coils so that their fields cancel in the center of the torus but cause the toroidal field lines to twist at the outside edge. A stellarator with two pairs of helical windings is known as $\ell = 2$ and the confined plasma has an elliptical cross-section. A stellarator with three pairs of helical windings is known as $\ell = 3$ and the plasma has a triangular cross-section. The classical stellarator is flexible for experiments because the toroidal and poloidal components of the magnetic field can be varied independently, but it is relatively difficult and expensive to build because the helical coils are trapped inside the toroidal coils and they have to be wound *in situ*. Variations of this basic stellarator concept have been developed (Figure 10.12) with the objective of simplifying the construction and optimizing the plasma confinement.

One important line of development was to dispense with the toroidal field coils by using a single set of helical coils with unidirectional currents to generate both the toroidal and the poloidal components of the magnetic field. This configuration (Figure 10.12a) is known as a *heliotron* in Japan and as a *torsatron* elsewhere in the world. It is easier to build than the classical stellarator because there are fewer coils and they are not trapped inside one another. Many heliotrons have been built in Japan and the most recent and largest is the *Large Helical Device (LHD)* at the National Institute for Fusion Science at Toki near to Nagoya. LHD started operation in 1998 and it has major radius $R = 3.7$ m and minor radius $a = 0.64$ m. A magnetic field of 3 T is produced by a single pair of superconducting helical coils ($\ell = 2$) that carry currents in the same direction. There is also a set of poloidal field coils to balance the vertical component of magnetic field produced by the helical coils. Steady-state operation of LHD has been demonstrated by sustaining a discharge for more than 1 hour. LHD uses neutral beam injection and ion cyclotron and electron cyclotron heating and has reached an electron temperature $T_e \approx 10$ keV, ion temperature $T_i \approx 13.6$ keV, and electron density $n_e \approx 1.0 \times 10^{21}$ m^{-3} (though not simultaneously). An important result is the achievement of high plasma pressure with $\beta \approx 5\%$ (albeit at low magnetic fields). These results show that LHD has plasma confinement properties comparable to those of the leading tokamaks of similar size.

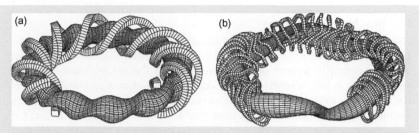

FIGURE 10.12 Coils and magnetic surfaces for (a) a *torsatron* or *heliotron* with $\ell = 2$ (there are two coils which carry currents in the same direction) and (b) an advanced modular stellarator (*helias*) with 5 periods each containing 10 nonplanar coils.

An alternative line of development dispenses with the helical coils and produces the twisted magnetic field by distorting the toroidal field coils into very complex nonplanar shapes (Figure 10.12b). Advances in the understanding of plasma confinement and in computing allow the coil design to be refined and effectively tailored to meet the plasma requirements. The *Wendelstein 7-AS* experiment, which operated at IPP Garching between 1988 and 2002, was the first test of this concept, which is known as a *helias*, for *helically advanced stellarator*. Confinement scaling was found to be similar to tokamaks and this success led to the larger *Wendelstein 7-X*, which, at the time of writing, is under construction at Greifswald in Northern Germany. Wendelstein 7-X will have 50 nonplanar superconducting coils, but the construction has a modular design so that there are only 5 different types of coil. Even with this simplification there have been many problems with construction, and the project, which was started in the mid-1990s, is not expected to be completed until 2014. It will have major radius $R = 5.5\,\text{m}$, average minor radius $a = 0.52\,\text{m}$, and toroidal magnetic field of $3\,\text{T}$.

had been developed, the only effective way to heat stellarator plasmas was by ohmic heating. This involved passing a current through the stellarator plasma (as in a tokamak), producing a poloidal magnetic field that competed with and degraded the inherent confinement properties of the fields produced by the helical coils. It was realized after many years of careful research that stellarators are very sensitive to errors in the magnetic fields (Box 10.6). They require very careful design and construction to have good confinement properties.

However, success with development of stellarator-like configurations in Japan and Germany has demonstrated plasma performance on a par with tokamaks. A variation of the stellarator design, with a simpler coil layout and known as the *Heliotron*, was developed in Japan during the 1970s and 1980s (Box 10.5). A machine of this type, the *Large Helical Device* (*LHD*) at the National Institute for Fusion Science at Toki near Nagoya, started operation in 1998 and has achieved impressive results. An alternative approach is the *Wendelstein* series of stellarators in Germany, which has developed the concept of twisted modular field coils (Box 10.5 and Figure 10.13). The first

Box 10.6 Magnetic Surfaces

Good plasma confinement in a closed toroidal system requires that the magnetic field lines form a set of *closed magnetic surfaces*. A useful way to visualize these magnetic surfaces is to think of the concentric layers that we see inside an onion when it is cut in half. A perfect magnetic-confinement system would be like a toroidal onion with toroidal magnetic surfaces nested one inside the other. A magnetic field line should always stay on the same magnetic surface as it spirals around the torus, and if we follow a field line many times around the torus it will effectively map out a magnetic surface. After the field line has made a single complete turn around the torus in the toroidal direction, it will have rotated in the poloidal direction. This is known as the *rotational transform angle*—denoted by the Greek letter ι (iota).

The rotational transform is related to the tokamak parameter known as the *safety factor q* (see Box 9.1) by the expression $q = 2\pi/\iota$. Thus, the $q = 3$ magnetic surface in a tokamak has a rotational transform angle $\iota = 2\pi/3 = 120°$.

The magnetic surfaces in a tokamak are inherently circular in their poloidal cross-section but they can be stretched vertically into *ellipses* and even made *D-shaped* by the external poloidal field coils. The *vertical elongation* and the *triangularity* introduced into the magnetic surfaces have important effects on the stability and confinement properties of the tokamak plasma. A tokamak is said to be *axisymmetric* because the magnetic surfaces have the same shape in each vertical plane all the way around the torus. In contrast, the magnetic surfaces in stellarators are *non-axisymmetric*, which means that the shape of the magnetic surface changes as one goes around the torus. For example, the magnetic surfaces in the $\ell = 2$ heliotron configuration (see Figure 10.12a) have an elliptical cross-section and the axis of the ellipse rotates with the helical coils when followed around the torus. The magnetic configuration has important effects on the plasma behavior and the theory of plasmas in non-axisymmetric systems (stellarators) is much more complex than in axisymmetric systems (tokamaks).

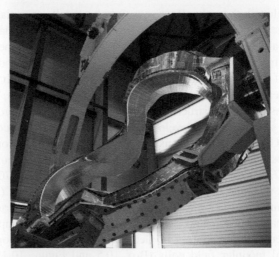

FIGURE 10.13 One of the nonplanar superconducting field coils for W7-X. The coil is shown supported by a yellow manipulator frame.

stellarator of this type was W7-AS, which operated at Garching from 1988 to 2002. Very encouraging results were obtained and led to the plans for a similar but much larger machine, W7-X, which is being built at Greifswald and is due to operate in 2014.

Conceivably these and other innovative ideas might have advantages over the tokamak and might find application in later generations of fusion power plants. But to wait for these concepts to reach the same stage of maturity as the line already explored by the big tokamaks would delay the development of fusion by many decades.

Stellarator of this type was W7-AS, which operated at Garching from 1988 to 2002. Very encouraging results were obtained and led to the plans for a similar but much larger machine, W7-X, which is being built at Greifswald and is due to operate in 2014.

Conceivably these and other innovative ideas might have advantages over the tokamak and might find application in later generations of fusion power plants. But to wait for these concepts to reach the same stage of maturity as the line already explored by the big tokamaks would delay the development of fusion by many decades.

Chapter 11

ITER

11.1 Historical Background

ITER was conceived as a collaborative venture between the United States, the Soviet Union, the European Union, and Japan following a political initiative between the Soviet and American leaders at the 1985 Geneva Summit Meeting. The principal objectives for ITER were to produce an ignited plasma and to study its physics, as well as to test and to demonstrate the technologies essential for building a fusion power plant—especially the superconducting magnets and tritium breeding systems. Although the four partners were supportive of international collaboration in fusion energy research, they were reluctant to commit themselves irrevocably to building the large ITER project. Furthermore, they did not want to decide immediately where it would be built. It was agreed to proceed in phases—conceptual design, engineering design, site selection, construction, and operation—but each phase would have to be approved without commitment to further phases. Each approval point has resulted in significant delays before agreement could be arrived at to proceed with the next phase, so the ITER project has taken much longer than originally expected.

Work on the conceptual design phase started in 1988, and a central ITER office was established in Garching (near Munich) in Germany. The conceptual design study, which established the key parameters and requirements for ITER, was conducted mainly as a part-time collaboration between scientists and engineers working in their home countries and was coordinated by round-table meetings in the central office. The successful outcome of the first phase was followed by an agreement signed in 1992 to proceed with the second phase—the Engineering Design Activity (EDA). This phase was scheduled to take 6 years and was carried out by a dedicated full-time "Joint Central Team" of scientists and engineers who were to be supported by many more experts working part-time in so-called "Home Teams." There was disagreement about where the central team should be based—Europe naturally thought that the ITER design efforts should continue in Garching—but the United States wanted the central team to move to San Diego and Japan wanted it to be based at Naka. Eventually, a compromise was reached to split the central team (which numbered about 160 professionals at its maximum) among all three places; this difficulty was an early sign of indecision at the political level that reccurred in years to come.

Fusion, Second Edition.

The engineering design of ITER went ahead with considerable enthusiasm and support from the international fusion community. Paul Rebut left his position as Director of JET and moved to San Diego to lead the engineering design, but he left the ITER project in 1994 and was replaced by Robert Aymar. Technical and scientific objectives were set and a detailed engineering design was prepared. The results from other tokamak experiments were now at the stage where it could be predicted with confidence that ITER would require a plasma current of 21 MA to reach ignition with an energy confinement time of about 6 s (Figure 10.8). The design for ITER was based on a tokamak with a divertor and a D-shaped cross-section—similar to JET but with physical dimensions about three times larger. The construction was planned to take about 10 years and was estimated to cost about six billion US dollars. These details were monitored and checked by independent experts, who agreed that ITER was well designed and could be built to schedule and within the estimated costs.

However, by 1998, when the engineering design had been completed and it was time to make the decision about the construction phase and to select a site where ITER would be built, the political climate had changed. The collapse of the Soviet Union and the end of the Cold War meant that a project like ITER could no longer be supported simply as a showcase of East-West collaboration—now it had to stand on its own merits as a new form of energy. Although there already had been considerable discussion of environmental issues during the 1990s, a consensus about the threat of global warming had been slow to find support in some counties. Alarm calls about the risk of future fuel shortages had generated little sense of urgency to carry out the basic research to develop alternative forms of energy. In fact, government funding for energy research, including fusion, had been allowed to fall in real terms. The situation reached a crisis point in the US, where a drastic cut in research funds caused the government to pull out of the ITER collaboration. The three remaining partners maintained their support for ITER in principle—but the Japanese economy, seemingly impregnable only a few years earlier, was in difficulties; the newly established Russian Federation (which had taken over the role of the former Soviet Union in the ITER partnership) was in economic crisis; and Europe did not have the determination to take the lead all by itself.

If anything was to be salvaged, it could be done only by accepting a delay until the political climate improved and by reducing the costs. The design team was given a mandate to redesign ITER at approximately half the cost of the 1998 design. It was clear that some reduction in performance would have to be accepted and that the goal of plasma ignition would have to be abandoned. The new goal was to be an energy gain factor $Q = 10$; with external heating power of 50 MW, the fusion reactions would produce an output of 500 MW. The revised design for ITER that was put forward in 2001 (see Table 11.1 and Figure 11.1) had physical dimensions about 75% of those of the previous design, with the plasma current reduced to 15 MA. The revised cost estimate came in (as instructed) at roughly half that of the previous design.

TABLE 11.1 ITER Parameters: The Main Parameters of ITER are Shown with Values for the Original Ignition Version Proposed in 1998, the Reduced Size ($Q = 10$) Version Proposed in 2001, and the 2011 Baseline Design.

Parameter		1998 Proposal	2001 Proposal	2011 Baseline	Units
Plasma major radius	R	8.14	6.2	6.2	m
Plasma minor radius	a	2.8	2.0	2.0	m
Plasma volume	V	2000	870	870	m^3
Toroidal magnetic field (at plasma axis)	B	5.68	5.3	5.3	T
Plasma current	I	21	15	15	MA
Safety factor	q	3	3	3	
Vertical elongation	κ	1.6	1.8	1.8	
Average plasma/ magnetic pressure	β	2.2–3.0	2.5	2.5	%
Normalized beta	β_N		1.77	2	
Average electron density	n_e	0.98	1	1	$10^{20}\,m^{-3}$
Average ion temperature	T_i	13	8	9.2	keV
Energy confinement time	τ_E	5.9	3.7	3.7	s
Energy multiplication factor	Q	Ignition	10	10	
Fusion power	P_F	1500	500	500	MW
Alpha particle heating	P_α	300	80	100	MW
External heating power at Q	P_{ext}	0	50	50	MW
Planned external heating power		100	73	73	MW

There was still no sign of a decision to start construction, but in 2001 the prospects that ITER would be built moved a step closer following an unexpected offer of a site near Toronto on the shore of Lake Ontario in Canada. One of the attractions of the offer was that Canada's existing nuclear fission power plants produce tritium as a by-product and this could be used as fuel for ITER. The Canadian initiative stimulated proposals for sites in Japan and Europe. Japan proposed a site at Rokkasho, north of Honshu, and Europe found itself with two contenders—a site at Cadarache near Aix-en-Provence in

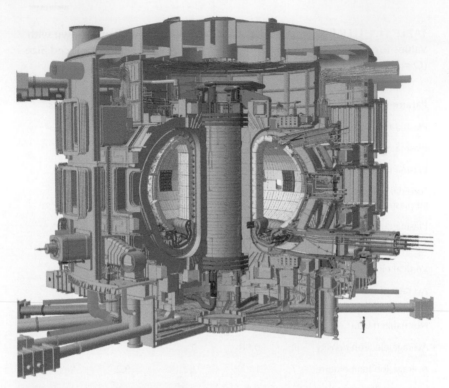

FIGURE 11.1 Cutaway model of ITER.

the South of France and one near Barcelona, Spain. Canada withdrew its offer
in 2003 and Europe eventually preferred Cadarache to Barcelona. However, a
decision between Cadarache and Rokkasho was more difficult as by now both
Europe and Japan were very keen to host ITER and neither would give way. It
took a further 2 years of hard negotiations to work out a compromise. It was
decided that ITER would be built in Europe at Cadarache but Japan would
have a share of the European industrial contracts. In addition, Europe agreed
to contribute substantially to the cost of other fusion activities in Japan as part
of a Broader Approach (see Box 13.6) to the final goal of fusion energy. These
include the engineering design and validation of an International Fusion Materials
Irradiation Facility (IFMIF), an International Fusion Energy Research Center
(IFERC), and an advanced superconducting tokamak at Naka (Figure 10.10). In
the meantime, the US had rejoined the project, China and South Korea had joined
as new partners, and India would join at the end of 2005, thus making seven partners. The agreed division of costs is that Europe will bear approximately 45%
of the overall cost of construction and the other six partners will each contribute
about 9%. In addition, France, as host country, will provide many of the support
facilities, including office buildings, improved road access, and power lines.

11.2   The Construction Phase Begins

The ITER Organization was established by an agreement signed by ministers of the seven members in November 2006 and was ratified by their governments the following year. This was good news, but 21 years had already elapsed since the project was conceived at the 1985 Geneva Summit. Japan nominated Kaname Ikeda as Director-General (he was a former diplomat). Norbert Holtkamp was appointed as the Principal Deputy Director-General and Leader of the Construction Project; he had an established reputation for building large accelerator projects to specification and cost but he had no previous fusion expertise. Further key appointments were made and the first ITER team members arrived at the Cadarache site, where they would be housed in temporary buildings for the next few years. They found themselves faced with a very difficult challenge—ITER was far from being as "ready to build" as some of the partners might have assumed. The engineering design was incomplete in some areas (the reduced-size ITER had been designed in the space of a year by a team that had been reduced in strength by the premature withdrawal of the US in 1999). Since 2001, the team had been further weakened, so that the design had not been fully updated to keep pace with ongoing developments in fusion research, and the cost estimates were incomplete in some areas.

The most urgent task of the newly-appointed team was to review the existing 2001 design, to identify any shortcomings, and to look at possible improvements and changes that might be required to make it consistent with more recent developments in fusion science. The principal goal of the review was to create a new Baseline Design that would confirm or redefine the physics basis and the requirements for the project; to confirm or alter the design of the major machine components and thus provide the basis for the procurement (in particular, for the long-lead items, including the vacuum vessel, magnets, buildings, etc.); and finally to provide input for the Preliminary Safety Report that was required in order to obtain approval from the French authorities. The ITER Design Review, which started in late 2006 and took a year to complete, was carried out by working groups formed by ITER team members assisted by many experts from the worldwide fusion community. It brought together more than 150 people, drawn from all the ITER members, to review the ITER design and issues, and to apply the latest state-of-the-art knowledge to its subsystems. A special emphasis was placed on areas of technical risk that would have major impact on the construction schedule and cost. It addressed the site-specific requirements and paved the way for a smooth licensing process. It also played a key role in making experts from the international fusion community feel that they are involved in ITER. The physics requirements were reviewed and updated—confirming ITER's goal to reach $Q = 10$ (see Box 11.1). Some of the important changes that resulted from the design review are discussed in more detail in Box 11.2.

Box 11.1　ITER Operating Space

When ITER comes into operation, a major effort will go into optimizing the plasma and machine conditions in order to obtain the very best performance. This process will take several years of close collaboration between physicists and engineers and will be built upon many decades of expertise developed on other tokamaks. Optimizing performance is a complex process and involves carefully balancing many factors. Some are determined by constraints set by the machine itself (examples are the maximum magnetic fields and the additional plasma heating power) and others are set by plasma operating limits (the maximum plasma current, density, and beta, as explained in Box 10.1). A full discussion would go well beyond the scope of this book, but a simplified overview can be gained if we look at the range of plasma density and temperature over which ITER can operate.

This *operating space*, calculated for a plasma current of 15 MA and a toroidal field of 5.3 T, is shown in Figure 11.2, where the electron temperature (in keV) is plotted against the electron density (in electrons per cubic meter). Both quantities are averaged over the plasma volume, and the peak values in the core of the

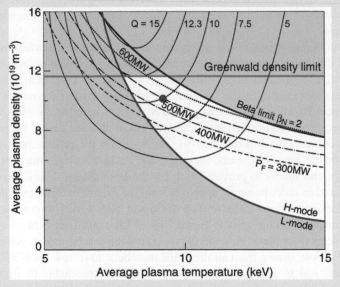

FIGURE 11.2　The ITER operating space, calculated for plasma current of 15 MA and toroidal field 5.3 T, where the average electron temperature (in keV) is plotted along the horizontal axis and the average electron density (in electrons per cubic meter) is along the vertical axis. The operating space lies within the non-shaded region, which is below the Greenwald density limit (the horizontal green line); in the H-mode regime (above the red line); and below the beta limit at $\beta_N = 2$ (the blue line). The black curves show values of Q and the dashed lines show fusion power ($P_F = 300$, 400, 500, and 600 MW). The baseline operating point at $Q = 10$ and $P_F = 500$ is marked by the red dot at 9.2 keV and $1 \times 10^{20}\,\mathrm{m}^{-3}$.

plasma will be higher than these average values. Most of this space is not accessible due to various constraints.

First we have to draw a horizontal line at the maximum average electron density that ITER will be able to reach—this is known as the *Greenwald density limit* (see Box 10.1), and it is predicted to be about $11.5 \times 10^{19} \, \text{m}^{-3}$. At densities above this line, the ITER plasma will be unstable and may disrupt. Then we draw two more constraining lines that start at the top of the figure and curve down toward the right-hand corner. The lower line is the limit between the L-mode and the H-mode—ITER has to operate to the right of (i.e., above) this line to be in the H-mode regime and to have good energy confinement. The second line is the *beta limit*, corresponding to $\beta_N = 2$ (see Box 10.1). ITER has to operate below this limit in the area below this line. This leaves a relatively small operating space—below the density limit and between the two other lines—that is marked as the unshaded region in the figure.

We can superimpose a series of dashed lines that represent the fusion power output in DT plasma (in the range 300 to 600 MW) and a series of solid curves that show the corresponding values of the fusion parameter Q (this is the ratio of fusion power to external heating power). ITER's goal of $Q = 10$ with 500 MW of fusion power would be reached with 50 MW of external heating power at an operating point that is marked with a red dot in the figure. The average plasma temperature would be about 9.2 keV and the average electron density would be about $10 \times 10^{19} \, \text{m}^{-3}$, corresponding to about 87% of the Greenwald limit. The energy confinement time $\tau_E = 3.8 \, \text{s}$ (calculated using the empirical scaling laws discussed in Box 10.4). The experience in present-day tokamaks is that energy confinement would deteriorate at densities closer to the Greenwald limit.

Using this type of approach, we can also look at the impact on performance if ITER is able to exceed the design values of plasma current and magnetic field or if, on the other hand, it is not able to reach them. The key parameter is the plasma current, and this depends on the toroidal magnetic field (see Box 10.1)—so the maximum current will be limited if ITER's magnets cannot run reliably at full field. Calculations show that a 10% reduction in the maximum field (to 4.8 T), with a corresponding reduction in the maximum current (to 13.5 MA), would bring Q down to about 6—but Q would go up to 20 if the current could be pushed up to 17 MA.

Box 11.2 Some Changes Resulting from the 2007 Design Review

Poloidal field requirements—The combination of high current, high fusion power, and long-pulse operation in ITER imposes very stringent demands on the poloidal field system to provide adequate inductive current drive and to control plasma shape and position. The vertical position of the plasma, the position of the divertor strike points, and the distance to the first wall in the presence of disturbances must all be controlled to small tolerances. Since the original design of the poloidal field coils was completed in 2001, studies of ITER simulation discharges have been carried out on JET, ASDEX-U, and DIII-D, leading to significant improvements in the understanding of the requirements for controlling high-performance discharges in ITER. The design review concluded that the design of the poloidal coils should be modified to upgrade the current in the coils and thus the capability of the poloidal field.

Vertical stability—A loss of vertical plasma position control in ITER will cause large thermal loads on plasma-facing components and also generate high electromagnetic loads. Further experiments on existing machines showed that the control parameter needed to be upgraded. To bring the ITER system up to the new standard, another set of in-vessel coils is required, with appropriate power supplies and connections.

Edge-localized modes—Edge-localized modes (known as ELMs) are MHD instabilities that destroy the magnetic confinement near the plasma boundary (see Box 9.9). They are driven by the large pressure gradients and current densities associated with the transport barrier near the plasma boundary that is responsible for the H-mode. Extensive studies in other tokamaks have shown that the ELMs can result in large energy deposition on the divertor target and other plasma-facing components on very short time scales, possibly resulting in extensive damage. The studies indicate that it is necessary to reduce the ELM energy by at least a factor of 20. ELMs tend to occur at random, but the energy deposition can be significantly reduced by inducing them at frequent, well-defined intervals so that single ELMs stay below a safe energy level. One way of doing this is by applying a field perturbation near the edge of the plasma. New coils applying a *resonant magnetic perturbation (RMP)* will be installed to do this. The RMP coils have to be aligned with the pitch of the equilibrium field and must be as close as possible to the plasma, in order to maximize the edge perturbation while minimizing the core perturbation.

Toroidal field ripple—Due to the finite number of toroidal field coils, there is a variation in the magnetic field from the positions under the coils to the positions between the coils. This ripple effect is most severe at the outer mid plane. Recent experiments on JT60-U and on JET have shown that, in H-mode, increasing the ripple can reduce the confinement of the thermal plasma by as much as 20%. Introducing ferritic steel inserts between the coils has been shown to reduce the ripple on the JT60-U tokamak, and it has been decided to use this method on ITER. One limitation is that the correction of ripple is exact only at one value of toroidal field, and, if the inserts are designed to minimize the ripple at the full field (5.3 T), the ripple will not be fully compensated during operation at other values of the field. Studies of this effect are continuing.

Plasma-facing materials—The constraints on plasma-facing materials are significantly more stringent for ITER than for smaller plasma experiments. In addition to coping with the high heat loads over a much more extended time than in present experiments, they have to minimize the effect of impurity contamination on plasma performance and operation, provide maximum operational flexibility, and minimize tritium retention during operation in the DT phase. Following the design review, it was decided that only two materials, beryllium and tungsten, should be used in ITER in the DT stage in order to avoid the long-term retention of tritium that has been found to occur in carbon.

11.3 Overview of the ITER Tokamak

ITER is based on the tokamak configuration of magnetic confinement discussed extensively in Chapters 9 and 10—but ITER will be much larger than any previous tokamak and its design and construction pose many new challenges. To minimize energy consumption, ITER uses superconducting magnets that lose their resistance when cooled with liquid helium to a temperature in the range of 4 K (minus 269°C). The toroidal magnetic field system has 18 coils (Figure 11.3), and each coil is wound from cables that contain a special superconducting alloy of niobium and tin that can operate at both high magnetic fields and high current. The cables are approximately 50 mm in diameter and consist of a mixture of copper and superconducting strands that are twisted together in order to avoid induced currents by changing magnetic fields. The copper strands are needed to carry the current if the superconductivity is lost (quenched) and the superconductor becomes resistive. The cable is contained in a steel case that provides the necessary structural support against the large electromagnetic forces; the current flowing through the cables is 60 kA and each coil experiences an inward radial force of 40,000 tons. The total weight of the toroidal field coil system will be over 6000 tons.

The toroidal field coils are wedged together in the core of the tokamak, leaving a vertical cylindrical space for the central solenoid. The central solenoid is made of six independent coil packs and also uses the high-field niobium–tin superconducting alloy strands. The main function of the central solenoid is to drive the inductive current in the plasma, and it also contributes to the poloidal magnetic field that determines the plasma shape, position,

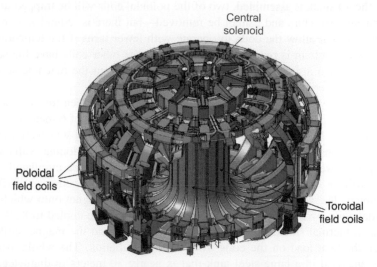

FIGURE 11.3 Overview of the toroidal and poloidal field magnetic coils.

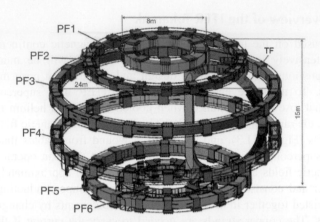

FIGURE 11.4   The poloidal field coil system, with a single toroidal field coil (TF) shown for reference. The central solenoid is not shown.

and stability. There are additional coils in the poloidal field system, including six very large horizontal coils that sit outside the toroidal coil structure, Figure 11.4. These coils operate at a lower magnetic field than the toroidal and central solenoid coils and they can use a superconducting alloy of niobium and titanium. Some of these poloidal coils are so large that they cannot be transported to the ITER site, so they will have to be wound at Cadarache in a specially-constructed building. There are some additional control coils that provide fine-tuning of the magnetic fields and allow compensation for various error fields that can arise due to small inaccuracies in positioning the big coils. After the tokamak is assembled, two of the poloidal coils will be trapped under the machine structure and cannot be removed—but there is redundancy in the design that will allow the coils to operate with fewer turns if the temperature of the liquid helium is reduced to 3.6 K. All of the other coils have the same redundancy and also are designed to be removable from the machine so that they could be replaced by new coils in case of a major fault.

The ITER vacuum vessel will be twice as large and sixteen times as heavy as that in any previous tokamak. With an internal diameter of 6 meters, it will measure over 19 meters horizontally by 11 meters high and will weigh more than 5000 tons (Figure 11.5). The steel vessel will have double walls with passages between them through which cooling water will be circulated. The vessel will be built in nine sections and will be welded together as ITER is assembled. The vacuum vessel sits inside the toroidal magnet onto which the poloidal coils are mounted. There is a thin metal structure cooled to 80 K that acts as a thermal shield between the vacuum vessel and the magnet coils to reduce the heat load on the coils from thermal radiation. The whole assembly is enclosed in a large steel tank that is nearly 30 meters in diameter and 30 meters high. This is the cryostat, out of which all the air is evacuated to

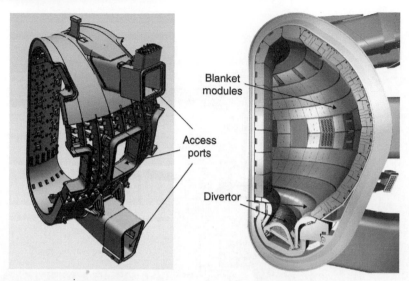

FIGURE 11.5   The ITER vacuum vessel, showing the blanket modules and the divertor.

provide thermal insulation for the superconducting magnets and other components. The cryostat has many external openings—some as large as 4 meters in diameter—which are connected by vacuum-tight tunnels to the vacuum vessel ports (Figure 11.6). The ports provide access to the vacuum vessel for plasma heating systems, diagnostics, and cooling systems. Some of the ports will be used for prototype blanket modules that will test the technology of tritium breeding required in a fusion power plant. During tokamak operation, the ports into the vacuum vessel are sealed off, but when the system is open for maintenance, the remote handling tools will pass through the tunnels into the interior of the vacuum vessel in order to service and remove the blanket modules and divertor components. Large bellows are used between the cryostat and the vacuum vessel to allow for thermal contraction and expansion between the two structures.

The inner wall of the vacuum vessel will be covered with the blanket modules. These will shield the superconducting magnets and the vacuum vessel itself from the high-energy neutrons produced by the fusion reactions. The neutrons are slowed down in the blanket, where their kinetic energy is transformed into heat energy—in a fusion power plant, this energy would be used for electrical power production, but in ITER it will be removed by water circulating through the blanket. The blanket modules contain about 80% steel and 20% water to optimize the slowing down and capture of the neutrons. The blanket has over 400 separate modules attached to the vacuum vessel and is designed so that individual modules can be removed by remote handling tools when maintenance is required. The front surface of a module can be detached from the body of the

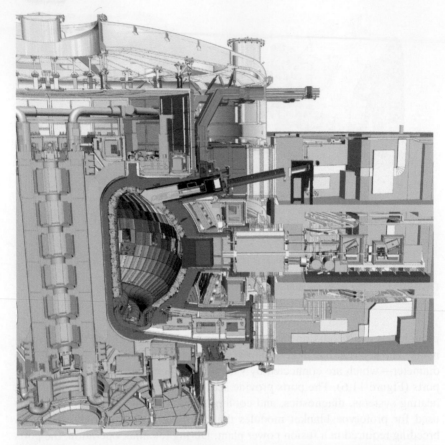

FIGURE 11.6 A model of a sector of the vacuum vessel and the cryostat showing the access ports. During ITER operation, many of these ports will contain diagnostic equipment (as shown here) and plasma heating systems. During maintenance periods, the ports will be used to admit the robotic arms for remote handling (see Figure 11.10).

module—this surface directly faces the plasma and has to remove the plasma heat flux. This surface will be covered with beryllium, a material that has good thermal properties and that minimizes impurity contamination of the plasma.

The divertor is located at the bottom of the vacuum vessel and is made up of separate "cassettes" that can be removed for repair and maintenance (Figure 11.7). Each cassette has three plasma-facing zones (usually known as targets). Two of these zones are located at the points where magnetic field lines intersect the divertor (see Figure 9.5). Ions and electrons from the plasma boundary layer are guided along these field lines and deposit their kinetic energy onto the targets—the heat flux is intense and the targets require special materials and active water cooling. There are few suitable materials for targets and testing suitable targets will be an important part of the ITER program. Carbon

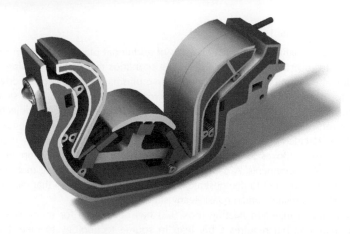

FIGURE 11.7 A model of a divertor cassette, showing the target surfaces that will be faced with tungsten tiles.

and tungsten are suitable target materials, but tungsten is preferred because it retains less tritium than carbon. The divertor is one of the most critical components in ITER and its success will be important for reliable operation.

ITER is the first fusion machine where the dominant plasma heating will be by alpha particles. Operating in DT plasma at $Q = 10$ with fusion power of 500 MW and 50 MW of additional plasma heating will correspond to total plasma core heating of 150 MW, of which 100 MW will be by alpha particles (400 MW of the fusion power is taken out of the plasma by the neutrons). However, ITER needs to have more than 50 MW of additional plasma heating to be able operate for the first few years in hydrogen and deuterium plasmas without tritium, when there will be no alpha particle heating. The present specification is for ITER to start operations with 73 MW of additional plasma heating, consisting of 33 MW of neutral beam injection, 20 MW of ion cyclotron heating, and 20 MW of electron cyclotron heating. Various options are being studied where the additional plasma heating could be increased up to 130 MW at a later stage. A mixture of different types of additional plasma heating is required (Box 11.3) in order to have the flexibility to operate ITER over a wide range of conditions and to be able to drive non-inductive plasma currents.

An extensive diagnostic system will be installed on ITER to provide the measurements necessary to monitor, control, and optimize plasma performance and to further the understanding of the plasma physics. These include measurements of plasma temperature, density, impurity concentrations and particle and energy confinement times, as well as monitoring the surface temperatures and behavior of the in-vessel components, such as the divertor. About 50 individual measuring systems will be incorporated, drawn from the full range of modern plasma diagnostic techniques, including lasers, X-rays, neutron cameras, impurity monitors, particle spectrometers, radiation bolometers, pressure and gas

Box 11.3   Additional Heating

ITER requires a mixture of different methods of additional plasma heating in order to be able to meet a wide range of operating scenarios that include heating and controlling the plasma and driving non-inductive plasma currents. Current drive is needed for plasma control and for the long-term goal of demonstrating a steady-state tokamak. It is planned initially to have a mixture of neutral beam injection, ion cyclotron heating, and electron cyclotron heating, with lower hybrid left as an option to be decided at a later stage. Although these techniques are widely used in existing tokamaks (Boxes 9.7 and 9.8), they all require extensive development to meet the ITER requirements. A common problem with all the heating methods is the need to develop and to demonstrate heating source technology that can operate routinely and reliably under quasi steady-state conditions.

Neutral beam injection heating (Box 9.7) has been used for many years in existing tokamaks but requires a big leap in source technology to meet ITER's requirements. The injection energy specified for ITER is 1 MeV, to ensure adequate beam penetration into high-density DT plasmas and in order to allow current drive in the plasma core. This high injection energy requires a negative deuterium ion beam source that is a major development from existing neutral beam source technology, which uses positive ions at much lower energies. ITER requires two neutral beam lines, each with 16.5 MW, giving a total of 33 MW (Figure 11.8). An option for a future upgrade would be to add a third beam line, increasing the total neutral beam power to 50 MW. In order to achieve 16.5 MW

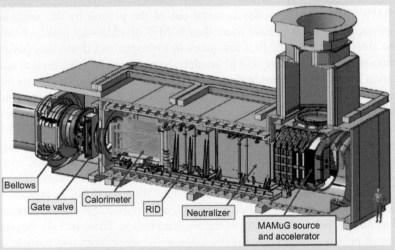

FIGURE 11.8   Schematic of a neutral beam injection heating system (see also Figure 9.7). Negative deuterium ions are produced in the ion source, accelerated to 1 MeV, and then neutralized. Any ions that have not been neutralized are removed by the residual ion dump (RID) and the powerful neutral beam (16.5 MW) passes into the ITER vacuum vessel (on the left-hand side of the figure).

heating in the plasma, an ion source delivering 40 A at 1 MeV is required. The best performance to date with neutral beam injection based on a deuterium negative ion beam source has been in JT-60U, which has achieved about 5 MW at 400 keV but only for a few seconds. Scaling up to 1 MeV, 16.5 MW, and very long pulses is a major challenge. An essential step to reaching the ITER requirements will be the construction of a test-stand for a full-size negative ion source and the full-scale prototype 1 MeV neutral beam system that is proposed in Padua.

Electron cyclotron heating is required for plasma heating, current drive in the plasma core, control of MHD instabilities, and plasma start-up. ITER has specified 20 MW of electron cyclotron heating at 170 GHz. The best performance to date in fusion experiments at this high frequency has been 3.6 MW for 2 s—longer pulses (about 60 min) have been reached only at the much lower power of about 100 kW. The 2007 design review concluded that electron cyclotron heating is one of the least risky options—but the development program needs to demonstrate that the 170 GHz gyrotron sources can run reliably at high power under steady-state conditions. The present specification is based on arrays of 1 MW gyrotrons. Prototype gyrotrons that are being developed in Japan and Russia are (at the time of writing) already close to meeting the ITER requirements. Europe is developing 2 MW gyrotrons that could be incorporated at a later stage.

Ion cyclotron heating is required for localized ion heating and central current drive, and to control sawteeth (Box 9.2). The ITER specification calls for steady-state operation with 20 MW in the frequency range 40 to 55 MHz. There is experience in this frequency range at MW power levels.

Lower hybrid is not included in the present specification for ITER but remains a possible option for future machine enhancements. It could be required for off-axis current drive in steady-state tokamak operation and for advanced tokamak operating scenarios aimed at improving the performance of ITER. A complication for both ion cyclotron and lower hybrid is that they require antennas inside the vacuum vessel and the coupling efficiency of the heating waves into the plasma depends sensitively on the plasma parameters at the plasma edge. This is a particular problem in H-mode regimes with ELMs and, in spite of recent progress to address this problem in other tokamaks, there still may be uncertainties on the quality of the power coupling in ITER.

There are options for future upgrading of the additional heating power from 73 MW to 130 MW with various combinations of the different methods, including increasing the neutral beam power from 33 to 50 MW, increasing the electron cyclotron heating from 20 to 40 MW, increasing the ion cyclotron from 20 to 40 MW, and adding 20 MW of lower hybrid.

analysis, and optical fibers. The arrangement of some of these systems is shown in Figure 11.9. Because of the harsh environment inside the vacuum vessel, these systems will have to cope with conditions not previously encountered in diagnostic implementation. The levels of neutral particle flux, neutron flux, and

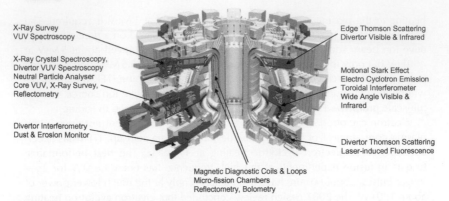

X-Ray Survey
VUV Spectroscopy

X-Ray Crystal Spectroscopy,
Divertor VUV Spectroscopy
Neutral Particle Analyser
Core VUV, X-Ray Survey,
Reflectometry

Divertor Interferometry
Dust & Erosion Monitor

Edge Thomson Scattering
Divertor Visible & Infrared

Motional Stark Effect
Electro Cyclotron Emission
Toroidal Interferometer
Wide Angle Visible &
Infrared

Divertor Thomson Scattering
Laser-induced Fluorescence

Magnetic Diagnostic Coils & Loops
Micro-fission Chambers
Reflectometry, Bolometry

FIGURE 11.9   Schematic of some of the diagnostics being built for ITER.

neutron fluence will be respectively about 5, 10, and 10,000 times higher than the conditions experienced in today's machines.

There are many very important systems external to the tokamak itself. ITER is the first fusion machine that is designed for continuous operation with deuterium and tritium fuels, and they will be processed in closed cycle. Less than 1 g of fuel will be present in the vacuum vessel at any one moment and it will be introduced as gas and as high-velocity pellets of solid DT ice. The gas refueling is most effective at the edge of the plasma, whereas the pellet refueling can be used to control the density in the core regions. The fuelling rate and response time, together with the correct isotopic mix, have to be carefully controlled to maintain fusion power at the required level. Only a small fraction of the fuel is consumed in the fusion reaction in each "pass" through the machine. The unused fuel is removed by the vacuum vessel pumping system and is processed through an isotope separation system that separates the hydrogen, deuterium, and tritium into separate storage systems so that the fuel can be used again.

The volume of the torus vacuum vessel is 1400 cubic meters, and that of the cryostat vacuum is 8500 cubic meters. Powerful vacuum pumping systems are required for the initial evacuation of all the air and to maintain the very low pressures that are required. Cryopumps cooled by liquid helium will be used for the neutral beam injection systems and for the cryostat surrounding the superconducting magnets. The cryoplant is composed of helium and nitrogen refrigerators combined with an 80 K helium gas loop. Three helium refrigerators will supply the required cooling to the cryo-distribution system. One of the helium refrigerators is fully dedicated to cooling of the cryopumps prior to cooling the cryostat. The other two refrigerators are used for the magnet system and provide partial redundancy for standby operation or during fault scenarios.

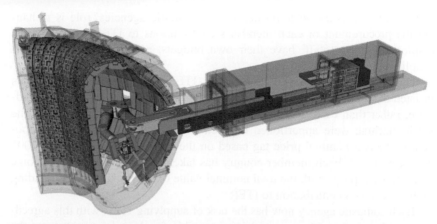

FIGURE 11.10   A schematic of the robot arm that will be used to maintain components inside the vacuum vessel.

Remote handling will have an important role to play in ITER. After operation in tritium is begun (presently planned for 2027), it will be impossible to make changes, to conduct inspections, or to repair any of the components in the activated areas other than by remote handling. Very reliable and robust remote handling techniques will be necessary to manipulate and exchange components weighing up to 50 tons. The reliability of these techniques will also influence the length of the machine's shut-down phases. All remote handling techniques developed for ITER operate on the same principle. A remote manipulator is used to detach the component (Figure 11.10); the component is removed through a port and placed into a docked transport cask; a temporary door is placed over the vacuum vessel access port; and the cask is closed to prevent contamination. The cask is moved on air bearings along to the Hot Cell. A similar docking occurs at the Hot Cell and the component is removed to be repaired or replaced. The process is then reversed to bring that component back to the vacuum vessel. The Hot Cell facility will be necessary at ITER to provide a secure environment for the processing, repair or refurbishment, testing, and disposal of activated components. Although no radioactive products are produced by the fusion reaction itself, energetic neutrons interacting with the walls of the vacuum vessel will activate these materials over time. Also, materials can become contaminated by tritium and by beryllium dust.

11.4   The Construction Schedule

The central team of the ITER organization based at Cadarache is relatively small—it numbered about 500 staff and 400 contractors at the end of 2010. It is supported by large teams working in each member country in organizations that are known as *domestic agencies*. The domestic agencies' role is to handle the procurement of each member's contributions to ITER—the agencies employ their own staff, have their own budgets, and place contracts with suppliers.

During the negotiations to set up the ITER organization, the members insisted that their contributions to ITER would be mainly in the form of hardware, rather than cash. Responsibilities for all of the component parts of the ITER machine were apportioned between the ITER members. Each component part has a nominal price tag based on the original estimates in the 2001 design proposal. Each member country has taken responsibility for a package of component parts, with the total nominal value of the package corresponding to that member's contribution to ITER.

Each domestic agency now has the task of supplying ITER with this agreed-upon package of components and is responsible for doing so at whatever the actual costs turn out to be. The ITER organization remains responsible for the overall design and specification of the machine and its integration, and for giving the domestic agencies the technical specifications for the individual components—but each domestic agency is responsible for procuring and for paying for these components. The level of technical specification varies—for some critical components, the ITER organization provides the domestic agency with a fully detailed design that is ready to go out for manufacture; in other cases, there is a technical specification but the domestic agency is responsible for completing the detailed design before it goes out to be manufactured.

This system does cause problems, because in the end someone has to pay for the real cost and this could have an impact on budgets for fusion research in some countries. There are also problems with matching priorities for components, as seen from the overview of the central organization, against annual budgets and cash flows allocated by the members to their domestic agencies. It would have been much simpler and more cost-effective to have funded the ITER construction centrally, but this arrangement was politically unacceptable to some of the members, who wanted to be sure that their own industries would benefit directly from ITER contracts.

France contributes to the ITER project as a member of the European Union. In addition, as host country for the project, France has undertaken a number of specific commitments, including providing a site of 180 hectares for the project and for carrying out all preparatory work, including clearing, leveling, and fencing the site and installing networks for water and electricity. France has established an international school for the families of ITER employees, has adapted the roads in the area for the transport of ITER components, and is constructing the ITER headquarters buildings into which the staff will move from their present temporary offices. France will also have the responsibility for dismantling and decommissioning the site at the end of ITER operations.

In mid-2010, the ITER Council approved the baseline cost, scope, and schedule and made some changes in the senior management of the ITER team.

FIGURE 11.11 (a) A model of the ITER site showing the large building in which the tokamak assembly will be installed and some of the ancillary buildings; (b) a photograph of the construction of the basement slab and seismic isolators (October 2011).

Norbert Holtkamp left the project and Osamu Motojima, who had extensive experience in fusion research as former director of Japan's National Institute for Fusion Science, was appointed to succeed Kaname Ikeda as Director-General. Figure 11.11 shows a model of the ITER site, with the large building in which the tokamak assembly will be installed and some of the ancillary buildings, together with a photograph showing the status in October 2011 of the construction of the basement slab and seismic isolators. The present authorized schedule aims at completion of the construction in late 2019, followed by the first pump-down of the vacuum vessel and the integrated commissioning of the tokamak systems. This will be followed by several years of plasma operation in helium and hydrogen to progressively optimize systems and plasma performance without activating the machine. The first operation in tritium is planned to start in 2027. Even this revised schedule is thought by many to be too optimistic, and it may have to be extended to take account of various issues, including the delays anticipated as a result of the earthquake and tsunami in Japan in 2011.

Chapter 12

Large Inertial-Confinement Systems

12.1 Driver Energy

The early calculations on ICF (inertial-confinement fusion) by John Nuckolls in 1972 had estimated that ICF might be achieved with a driver energy as low as 1 kJ. More cautious estimates led to the construction of Livermore's Shiva laser, with 10 kJ. However, during the 1970s, estimates of the required driver energy increased steadily to values that were beyond the capability of Shiva or other foreseeable lasers. Moreover, the experiments on Shiva had shown that the infrared light output from a neodymium laser was not as effective in compressing a target as had been expected. The infrared did not penetrate deep enough into the surrounding plasma and it heated the electrons, which in turn heated the core prematurely, making it more difficult to compress. To make a more effective driver, it was necessary to convert the infrared into ultraviolet light or into X-rays, even though some of the driver energy would be lost in the conversion process.

In order to provide reliable experimental data on the minimum energy required for ignition, a series of secret experiments—known as *Halite* at Livermore and *Centurion* at Los Alamos—was carried out at the nuclear weapons test site in Nevada between 1978 and 1988. The experiments used small underground nuclear explosions to provide X-rays of sufficiently high intensity to implode ICF capsules, simulating the manner in which they would be compressed in a hohlraum. The results of these experiments have never been made public—indeed, there is very little open reference to the work at all beyond a brief statement, in a review published in 1995, that the underground tests "demonstrated excellent performance, putting to rest fundamental questions about the basic feasibility of achieving high gain." It has been revealed more recently that similar experiments had been carried out independently (and possibly earlier) at the Nevada test site by teams studying ICF in the UK—and it is reasonable to assume that similar tests have been carried out in the Soviet Union.

Although the detailed results of all of the experiments still remain secret, it seems that the Halite/Centurion results predicted values for the required laser energy in the range 20 to 100 MJ—higher than the predictions of theoretical models of capsule implosions and much larger than could be reached with existing lasers. The Nova laser that replaced Shiva at Livermore in the early 1980s could deliver about 40 kJ of ultraviolet light. This led to proposals in the US in the late 1980s to build a facility with a 10 MJ laser that could achieve fusion yields in the range 100 to 1000 MJ. Such a laser was thought to be technically feasible—but it would have been extremely expensive. After several years of debate and redesign, the proposal took final shape as the *National Ignition Facility (NIF)* at the Lawrence Livermore National Laboratory in California—a laser that would have a more modest output of 1.8 MJ of ultraviolet light but that was still very much larger than Nova.

A demonstration of ignition and burn propagation with indirect drive is the next important goal of ICF and is the objective of two very similar projects: NIF, which began construction in 1997 and was completed in 2009, and the Laser Mégajoule (LMJ), which is being built in France near Bordeaux, with completion planned for 2014. Experiments on NIF started in 2010, and its stated goal is to demonstrate ignition by the end of 2012. Both projects use multiple-beam neodymium-glass lasers (in fact NIF and LMJ share some of the laser technology) capable of delivering up to 1.8 MJ of ultraviolet light. Research in ICF is an important part of the scientific programs planned for both projects—although their main funding comes from the perceived need in the US and France to maintain scientific expertise in nuclear weapon design following the introduction in 1996 of the treaty banning all underground testing of nuclear weapons. The new laser facilities will not be able to test actual weapon systems, but they will allow weapon designers to continue to study and to maintain their expertise in the physics of matter compressed to very high densities. The military side of both projects remains highly secret, but the civil applications of the ICF research at NIF and LMJ are in the public domain, and it is these aspects that we discuss here.

12.2   The National Ignition Facility

NIF is the latest (and by far the largest) in a series of lasers built at Livermore to study ICF. The lasers use neodymium glass to produce light in the infrared region of the spectrum at 1.053 μm; the light is then converted using special nonlinear crystals into ultraviolet light with a wavelength of 351 nm (one third of the wavelength of the original infrared light). At the time of writing, NIF is the world's largest laser and can deliver much more energy than any previous ICF laser system. The original cost estimate of the NIF project was about $1 billion, but costs escalated to over $4 billion during construction, resulting in criticism of the project management by the US Congress.

NIF's laser (Figure 12.1) starts with a master oscillator, which generates a very-low-energy pulse with a few nanojoules (10^{-9} J) of high-quality infrared light at a wavelength of 1.053 μm. The time evolution and other characteristics can be tailored to meet the specific requirements of the experimental shot. The

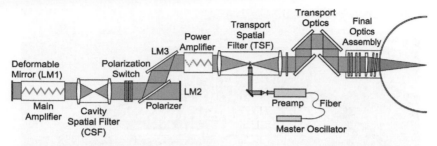

FIGURE 12.1 A schematic of one of the amplifiers of the National Ignition Facility (NIF) at the Lawrence Livermore National Laboratory. The main amplifier section has 11 large slabs of neodymium glass, and the power amplifier has five slabs that are pumped by arrays of flash lamps (not shown). Laser light enters from a master oscillator and passes four times through the main amplifier and twice through the power amplifier sections before being directed to the target chamber. The infrared light from the laser is converted into ultraviolet light in the final optics assembly before it enters the target chamber. NIF has 192 of these amplifiers in parallel to meet its requirement of 1.8 million joules.

low-energy light pulse is then split into 48 beams, which pass through 48 identical preamplifiers for initial amplification and beam conditioning. In the preamplifiers, the energy of each beam is increased more than a billion (10^9) times, up to a few joules. Each beam is then split again into four beams—making a total of 192 beams—and each one passes back and forth several times through its own series of main amplifiers (Figures 12.1 and 12.2). These contain large slabs of neodymium glass surrounded by xenon flashlamps that are energized electrically, creating an intense flash of white light; this is absorbed by the neodymium atoms—"pumping them up" to a higher energy state, which then is released as the short burst of laser light. After passing through the amplifiers, each laser beam has energy of about 20 kJ, and the combined energy of the 192 beams after leaving the main amplifiers is about 4 MJ of infrared light. This is a gain of one million billion (10^{15}) times the energy of the beam that started out in the master oscillator. The beams then pass through an array of mirrors where they are reconfigured into groups of four, so that there are 48 groups, each with four beams. The beams pass through the final optics assemblies positioned symmetrically around the target chamber to line-up the beams precisely on the target. Inside the optics assemblies, KDP (*potassium dihydrogen phosphate*) crystals convert the beams from infrared light into the desired ultraviolet light and other optics focus the light onto the target.

Since some energy is lost in the wavelength conversion, the final energy of the beams that can be focused onto the target is about 1.8 MJ. Delivered in a pulse that lasts about 15 nanoseconds, this corresponds to an instantaneous power of 500 terawatts (5×10^{14} W). To put these numbers in more familiar terms, 1.8 MJ is approximately the energy released by burning 135 g of coal or about 50 ml of gasoline. When this *energy* is released as a short laser pulse, the instantaneous *power* of 500 TW is equivalent to the combined output of

FIGURE 12.2   One of the two identical laser halls of the National Ignition Facility (NIF) at the Lawrence Livermore National Laboratory. Each hall has two clusters of 48 beam lines, one on either side of the utility spine running down the middle of the bay.

250,000 large (2 GW) coal-fired or nuclear power plants. When the beams arrive at the target, they have to be aligned to within 60 μm—a precision that is equivalent to throwing a coin from Livermore to San Francisco (a distance of about 64 km) and landing it perfectly inside the coin slot of a parking meter. All of this high-tech equipment is housed in a 10-story building that is the size of three football fields.

The beams enter the target chamber from 48 points symmetrically positioned around the chamber's top and bottom sections and are focused onto the ICF target. The tiny target assembly is positioned with an accuracy of less than the thickness of a human hair at the center of the target chamber, which has a diameter of 10 meters (Figure 12.3). NIF is designed primarily to use the *indirect drive* method of operation, in which, as discussed in Chapter 7, the ICF capsule is housed inside a small metal cylinder known as a *hohlraum*. A typical capsule (Figure 12.4b) is about the size of a grain of rice—a hollow sphere of silicon-doped plastic or beryllium–copper alloy with a diameter of 2 mm that is filled with a mixture of deuterium and tritium gas. The capsule is cooled to about 18 K (−255°C) and a very thin layer of solid DT ice forms on the inside wall. A typical hohlraum is a small gold cylinder with openings at both ends (Figure 12.4a). The laser beams are focused into these holes (Figure 12.4c) and they heat up the inside of the hohlraum, converting the

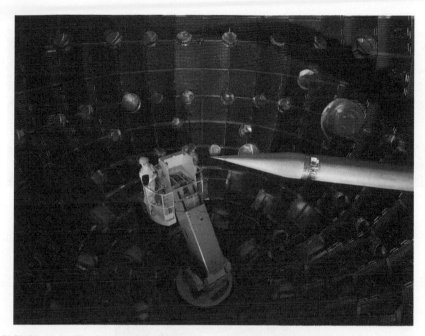

FIGURE 12.3 The interior of the NIF target chamber, showing the service module that allows technicians to inspect and maintain the chamber. The target is supported on the positioning arm, which is seen on the right.

energy of the ultraviolet light into X-rays (Figure 12.5). More advanced target designs are being developed and are discussed in Box 12.1.

The X-rays heat and vaporize the surface of the capsule and the rapidly escaping vapor pushes inward on the capsule, causing it to implode under a peak pressure in excess of 100 Mbar (100 million times the Earth's atmospheric pressure). As the fuel is compressed, it produces a shockwave that heats the fuel in its core to a temperature at which fusion reactions will occur—the core will ignite and the fusion energy released heats up the surrounding fuel to thermonuclear temperature. The entire process takes just 20 billionths of a second. More than a dozen laser parameters are adjusted to control and optimize key physics parameters. For example, in ignition experiments, the master oscillator must produce a pulse consisting of four shocks that are timed to collapse the capsule in a precise sequence. Sudden amplitude (peak power) transitions in the pulse create these shocks, and their timing must be exact to create the ignition "hot spot"—which starts the fusion burn—at the center of the compressed fuel. The amplitude, duration, timing, and energy of each shock can be manipulated by producing the desired pulse shape in the master oscillator of the laser system.

Of the 4 MJ of infrared laser energy, 1.8 MJ remains after conversion to ultraviolet, and about half of this remains as X-rays after conversion by the

(a)   (b)

(c)

FIGURE 12.4   Three views of the NIF target assembly: (a) the hohlraum; (b) the capsule, which is about 2 mm in diameter, filled with cryogenic hydrogen fuel; (c) the capsule seen through the laser entrance hole of the hohlraum.

hohlraum. Typically, 10–20% of the X-ray energy will be absorbed by the outer layers of the capsule and the shockwave created by this heating typically contains about 140 kJ. This compresses the fuel in the center of the target to a density that is about 10 times the normal density of lead. If everything works as planned and the capsule ignites, between 10 and 20 MJ of fusion energy will be released. In this case, the *fusion energy gain*—defined as the fusion energy output (10–20 MJ) divided by the laser driver energy (1.8 MJ)—will be between 5 and 10. In principle, improvements in both the laser system and hohlraum design could increase the shockwave energy to about 420 kJ and the fusion energy to about 100 MJ—but in practice experiments might have to stop at 45 MJ (equivalent to exploding about 11 kg of TNT) to avoid damage to the NIF target chamber. Even if the most optimistic fusion energy output of 45 MJ can be reached, it will be equivalent to only 10% of the input energy (about 400 MJ) required by the electrical system that drives the laser—so NIF will not be a net producer of energy. (We discuss the target and driver requirements for a fusion power plant in Chapter 13.)

The construction of NIF was completed early in 2009, when the final modules of the laser system were installed and all 192 laser beams were fired

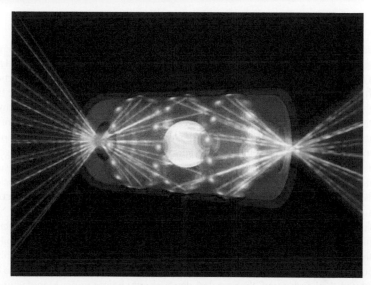

FIGURE 12.5 An artist's rendering of an NIF target pellet inside a hohlraum capsule, with laser beams entering through openings on either end. The beams produce X-rays which compress and heat the target to the necessary conditions for nuclear fusion to occur.

together into the target chamber. During 2009 and the first months of 2010, the many items of complex equipment that make up the NIF system were commissioned, tested, and calibrated in readiness for initial integrated ignition experiments. By the end of 2010, the laser beams had fired 1 MJ of ultraviolet energy into the first capsule and a shot into a capsule filled with deuterium and tritium gas had produced 3×10^{14} (300 million million) neutrons—a new record for the highest neutron yield by an ICF facility.

12.3 Laser Mégajoule (LMJ)

Laser Mégajoule (LMJ) is being built near Bordeaux by the French Atomic Energy Commission (CEA), and it is the largest ICF experiment to be built outside the US. LMJ has objectives and specifications very similar to those of its American counterpart, NIF, and is designed to deliver up to 1.8 MJ (550 TW) of ultraviolet laser light at 0.351 μm onto its target. Construction of the project is well advanced, with completion scheduled for 2014.

LMJ will have 240 identical laser beam lines arranged in 30 groups, each with eight beams. The beam lines are housed in four laser halls located on either side of the experimental hall containing the target chamber. As is the case in NIF, each beam line uses neodymium-glass amplifiers that are pumped using xenon flashlamps, and the beam makes four passes through each amplifier in order to maximize the energy gain. The target chamber consists of an aluminum sphere, 10 m in diameter, fitted with several hundred ports for the

Box 12.1 Advances in Target Design

In Box 7.4 we discuss the basic elements of target design. This is an area of intense development and some recent improvements to the design of the targets for LMJ and NIF could lead to more efficient performance and allow ignition to be reached with reduced driver energy. There is a major effort to develop new materials for the target components and to refine the existing ones. The hohlraum is made from a material with a high atomic number, such as gold or uranium, which is efficient at generating X-rays. Exotic mixtures of these elements are being tested—for example, changing the wall material in the hohlraum from pure gold to a mixture of gold and uranium reduces the required driver energy by 10%. In contrast, the capsule is made from light elements with low atomic numbers that perform well as "rocket fuels" when ablated by the X-rays produced in the hohlraum. Plastics (which contain mainly carbon and hydrogen) are traditionally employed, but more recent NIF designs have used beryllium or beryllium–copper alloys, which have the advantage of higher density and high thermal conductivity. The copper improves the ability of the target to absorb X-rays, and the mechanical strength of beryllium is high enough to contain the fuel as gas at room temperature. This allows the targets to be filled with fuel and stored, then cooled down shortly before firing. Manufacturing requirements for all ICF targets are extremely rigid and components have to be machined to an accuracy of less than one micrometer (10^{-6} m).

The reference target for LMJ is based on a cylindrical gold hohlraum that is 6.2 mm in diameter and 10.4 mm long, with walls 50 μm thick. It is designed to be used with all of LMJ's 240 laser beams, which will enter the hohlraum through 3.5-mm-diameter access holes at either end. These holes are covered with a thin polyimide film 1 μm thick, so that the hohlraum can be filled with a mixture of hydrogen and helium gas at a density of 0.83 mg cm^{-3}. The conversion of ultra-violet laser light into X-rays is relatively inefficient—out of 1.4 MJ of laser energy absorbed by the LMJ reference target, only 0.16 MJ (about 11%) is finally absorbed in the capsule, and 0.36 MJ (about 26%) escapes through the end holes. The capsule has a diameter of about 2 mm and consists of a plastic ablator shell (about 175 μm thick) that is doped with germanium and that has a layer of solid DT ice (about 100 μm thick) on the inside of the ablator shell. Targets with slightly different dimensions are being prepared for experiments with a smaller number of laser beams delivering reduced energy—for example, 160 beams delivering up to 1.4 MJ.

The traditional design for the hohlraum is a cylinder, although spherical hohlraums have been considered—but recently a novel design, shaped like a rugby ball, has been studied with theoretical modeling at LMJ and experimental testing on the OMEGA laser. The new shape appears to be more efficient because it minimizes the hohlraum wall surface (approximately half the laser energy is used to heat the walls) while keeping the equatorial radius constant.

injection of the laser beams and the introduction of diagnostics. The chamber (Figure 12.6) is designed to absorb up to 16 MJ of neutrons, 3 MJ of X-rays, and 3 MJ of particle debris. LMJ will use indirect drive and is developing some advanced targets (Box 12.1).

FIGURE 12.6   The LMJ target chamber in May 2010.

Construction of LMJ started with building a prototype of one of the 30 chains of eight beam lines. This is known as the *laser integration line (LIL)*. It was completed in 2002 and has been used to optimize the design, performance, and components of the full laser system. LIL delivers 30 kJ of ultraviolet light. It has its own target chamber, 4.5 m in diameter, equipped with a target and various diagnostic facilities that can be used for a wide range of studies of laser-plasma interactions. An interesting development is the construction alongside LIL of a new ultrafast pulse laser, known as PETAL, that will generate pulses with energy up to 3.5 kJ and a duration between 0.5 and 5 ps at a wavelength of 1.053 μm. PETAL is currently under construction and the first shot on target is planned for 2012. It will be linked to the beam from LIL to carry out fast-ignition experiments (see Boxes 7.5 and 12.3). This project, funded by the French region of Aquitaine, will also be used as a test bed for studying the physics and laser technology needed for the proposed European facility known as HiPER (see Section 12.6).

12.4 OMEGA and OMEGA EP

The OMEGA laser at the University of Rochester in upper New York State has been in operation since 1985. Originally having 24 beams delivering a total energy of 2 kJ of ultraviolet light, the facility was upgraded in 1995 to have 60 beams delivering 30 kJ in a pulse lasting 1 ns.

OMEGA has concentrated on the direct-drive approach to ICF, which puts stringent requirements on the uniformity with which the laser energy illuminates the fusion capsule. Each beam must be focused to a uniform spot and variations in power between beams must be minimized. A large number of beams improve the spatial uniformity of the target illumination because the beams overlap each other so that the energy falling on any point on the capsule surface is averaged over several beams. In this way, a requirement that the uniformity of the energy distribution on the capsule surface should be in less than 2% translates into a requirement that the variation in energy between individual laser beams should be less than about 4%. This is achieved by carefully balancing the gain produced by each amplifier and by making sure that all the beams arrive at the target at the same time. Heating the capsule ablates the surface and applies an inward pressure to compress the capsule—but it also generates shockwaves that propagate into the fuel and heat it. It is important to avoid preheating the fuel (compressing a hot gas requires more energy than compressing a cold one) and this means that only very weak shocks with strengths of less than a few million atmospheres (Mbar) can be tolerated at the start of the implosion. This conflicts with the requirement that pressures over 100 Mbar are needed to drive the implosion at a sufficiently high velocity (over 3×10^9 meters per second) to avoid the capsule breaking up due to the Rayleigh-Taylor instability. The required pressure increase from a few Mbar to over 100 Mbar has been achieved in OMEGA by launching the laser energy in a series of very short spikes (known as a *picket pulse*) to generate a sequence of shocks of increasing strength.

Recent results with direct drive on cryogenically-cooled capsules filled with DT obtained a parameter ρr (the product of the fuel density ρ multiplied by the capsule radius r, as explained in Box 7.1) equal to about $0.3 \, \mathrm{g \, cm^{-2}}$. This is the highest value of ρr yet obtained, corresponding to a peak fuel density of about $250 \, \mathrm{g \, cm^{-3}}$ (about 22 times the density of lead). The Lawson parameter $P\tau$ (*pressure* multiplied by *energy confinement time*) has been inferred from the measurements of ion temperature (about 2 keV), areal density ($0.3 \, \mathrm{g \, cm^{-2}}$) and neutron yield (about 5×10^{12} neutrons per second), giving a value of about 1 bar s.

An upgrade known as OMEGA EP, completed in 2008, has an additional four laser beam lines—two of which can produce the ultrashort picosecond pulses that are required for fast-ignition experiments. Generating such short-duration high-energy laser pulses is difficult because the intense high power would damage the amplifiers. To overcome this problem, OMEGA EP uses a smart technique known as *chirped-pulse amplification,* where the initial low-energy laser pulse is first made thousands of times longer, then this long pulse is amplified without damage to the amplifiers, and finally it is recompressed into a very short and very intense pulse (Box 12.2). The OMEGA EP laser intensity on target is expected to eventually reach $10^{21} \, \mathrm{W \, cm^{-2}}$, inducing an electric field so large that the electrons of the capsule will be accelerated to a velocity close to the speed of light.

Fast-ignition experiments on the OMEGA EP laser system have been carried out with deuterated plastic shell targets compressed with an initial energy of about 20 kJ before being heated with a second ultrashort pulse. The second

Box 12.2 Chirped-Pulse Amplification

Chirped-pulse amplification (CPA) is a technique for amplifying an *ultrashort laser* pulse up to the *petawatt* (10^{15}W) power level. Such very high powers cause serious damage to the amplifiers and other optical components used in lasers. The critical damage threshold in these components is reached when the power per unit area exceeds a few gigawatts per square centimeter (10^9W cm^{-2}). NIF delivers 500 tera-watts (5×10^{14}W) in a pulse that lasts a few nanoseconds (10^{-9}s) and stays below the critical damage threshold by using very large amplifiers with an enormous sur-face area. The problem becomes increasingly more serious with ultrashort-pulse lasers because instantaneous power increases as the pulse is shortened.

CPA gets around the problem by stretching out the pulse in time before it passes through the amplifiers and then shortening it again afterward. We usually refer to laser light as being monochromatic—this means that it consists of just one color at a precise wavelength. But an ultrashort pulse of laser light actually contains a small range of colors or wavelengths—albeit with a very narrow spec-trum when compared to non-laser light sources. The clever trick used with CPA is to pass the light before amplification through a pair of optical devices, known as *diffraction gratings*, that separate the light into its different wavelengths. The gratings are arranged so that the long-wavelength components of the laser pulse travel a shorter path than the short-wavelength components, so that the pulse is stretched out in time. This is known as a *positively chirped pulse*—the shorter wavelengths lag behind the longer wavelengths. The pulse duration of the chirped pulse can be made up to 100 thousand times longer than the original pulse and its intensity is reduced accordingly. The chirped pulse, whose intensity is now below the damage threshold of gigawatts per square centimeter, can be ampli-fied safely by a factor 10^6 or more. Finally, the amplified chirped pulse is recom-pressed back to the original pulse duration when it passes through a second pair

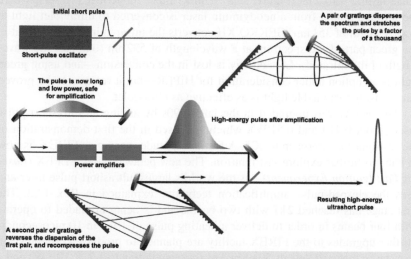

FIGURE 12.7 A schematic of the principle of chirped-pulse amplification.

of gratings arranged in the reverse order so that the shorter wavelengths catch up with the longer ones (see Figure 12.7).

The term "chirped pulse" comes from the characteristic chirp-like call of a bird—the sound pitch usually increases or decreases from start to finish. Chirped-pulse amplification was invented in the 1960s as a technique to increase the available power in radar systems and was applied to lasers in the mid-1980s. Some examples of ICF lasers that use the technique to study fast ignition are the Vulcan Petawatt Upgrade at the Rutherford Appleton Laboratory in the UK, the GEKKO petawatt laser at the GEKKO XII facility at Osaka University, the OMEGA EP laser at the University of Rochester, and the PETAL laser near Bordeaux. In addition to these applications in ICF, CPA makes it possible to build miniature lasers, leading to compact high-power systems known as "tabletop terawatt lasers."

pulse delivers energy of 1 kJ in a pulse lasting 10 ps, which corresponds to a power of about 10^{14} watts (100 million million). Initial results show that the neutron yield when the fast heating is applied is up to four times larger than the yield when a capsule is compressed without the fast-heating laser pulse. Encouraging preliminary results have also been achieved with a new concept known as *shock ignition* (Box 12.3).

12.5   FIREX

The GEKKO XII laser facility at the Institute of Laser Engineering at Osaka University in Japan has 12 beam lines that deliver a total of about 10 kJ in a pulse lasting 1 to 2 ns (Figure 12.8). Unlike other modern ICF lasers, where the infrared output from a neodymium laser is converted to ultraviolet light at a wavelength of 351 nm, GEKKO XII converts the infrared into visible light in the green part of the spectrum at a wavelength of 532 nm (one half the wavelength of the infrared). Less energy is lost in the conversion—and using green light is an option under consideration for HiPER—but it remains to be proved conclusively that visible light is as effective as ultraviolet.

The facility was upgraded in the late 1990s by adding an ultrashort-pulse laser (about 0.4 kJ and 0.5 PW), which was used in the first demonstration of the fast-ignition approach to ICF. A further upgrade, known as FIREX, is now in hand to further explore fast ignition. The new petawatt laser—LFEX (*Laser for Fast Ignition Experiment*)—is the world's largest ultrashort-pulse laser and uses the chirped-pulse amplification technique explained in Box 12.2. The first stage has reached 2 kJ with two beams and is being upgraded to operate with four beams in order to deliver a heating pulse of 10 kJ in 10 picoseconds. Further upgrades to the FIREX facility are planned to produce a compression laser pulse that will deliver 50 kJ in 3 ns (corresponding to 1.7×10^{13} W) and a heating laser pulse of 50 kJ delivered in 10 ps (5×10^{15} W). A new design for

Box 12.3 Approaches to Ignition

The basic concept of using laser light to compress and heat an ICF capsule to ignition (see Chapter 7) has evolved into several different approaches.

The original concept—known as *direct drive*—uses ultraviolet or visible light from a single laser focused directly onto the fusion capsule to compress the fuel and to heat the core to fusion temperatures. The surface of the capsule has to be very smooth and the laser light has to be very uniformly distributed over the surface in order to compress the capsule uniformly and to avoid instabilities that would break it up before it reaches sufficiently high density and temperature. The direct-drive route to ICF has been studied at the University of Rochester in the US using the OMEGA laser (60 beams delivering about 30 kJ of ultraviolet light onto the target) and at the University of Osaka in Japan using the GEKKO laser (12 beams delivering 15 kJ of green light).

The second approach—known as *indirect drive*—was initially developed in secret during the 1970s and 1980s at Livermore, where it was studied with the Nova laser (10 beams and 40 kJ of ultraviolet) and at Aldermaston in the UK, where it was studied using the two-beam HELEN laser. The fusion capsule is placed inside a small metal cylinder—known as a *hohlraum*—which is heated by the laser beams to produce X-rays that in turn compress and heat the capsule. Indirect drive is less efficient than direct drive because energy (typically about 50%) is lost in the conversion to X-rays, but the X-ray energy is more efficiently absorbed by the capsule and generates a higher ablation pressure with a more stable and more symmetrical implosion. The NIF and LMJ projects (1.8 MJ of ultraviolet) will concentrate on the indirect-drive approach. Both NIF and LMJ could be used to investigate direct drive but this would be very expensive using NIF if the arrangement of laser beams needs to be changed. On LMJ there is an interesting possibility to study direct drive using two out of the three rings of laser beams.

Direct and indirect drive aim at central isobaric ignition as a single-step process, with the driver energy carefully controlled so that the fuel is compressed with as little preheating as possible—until a small region in the core (the "hot spot") reaches a high temperature at the time of maximum compression and starts the fusion burn. In contrast, the next two approaches follow a two-step sequence where the fuel is first compressed with as little heating as possible—and then the compressed core is rapidly heated to start the fusion burn.

The approach known as *fast ignition* separates the compression and heating stages, using separate lasers for each stage. The key element is to use an ultrafast laser to heat the core of the fuel capsule after it has been compressed by the first laser. This concept has been tested using the GEKKO laser in Osaka and by OMEGA EP at Rochester. If this approach works when scaled up, it would require less driver energy than direct or indirect drive. The combined energy of the two lasers planned for HiPER (see Section 12.6) would be 270 kJ and the fusion yield would be 25 to 30 MJ—a gain of about 100.

A more recent approach, known as *shock heating*, also aims to separate the compression and heating but in this case would use a single laser that would first compress the fuel capsule without heating it. Then the same laser would launch a strong, spherically convergent shockwave into the compressed fuel. The ignition shock is timed to collide with the return shock near the inner shell surface. After the ignition and return shock collide, a third shock propagates inward, leading to further compression of the hot spot. The final fuel assembly develops a centrally peaked profile. Such non-isobaric assemblies exhibit a lower threshold than standard isobaric ones. Shock heating would have the advantage of reducing the energy required for ignition using a single laser system, whereas fast ignition requires two separate laser systems.

FIGURE 12.8   The 12-beam GEKKO laser, at Osaka University's Institute for Laser Engineering in Japan, was completed in 1983. GEKKO is used for high-energy density physics and inertial-confinement fusion research.

the fast-ignition target (Figure 12.9) is being developed as part of the FIREX program.

12.6   HiPER

HiPER (High Power laser Energy Research) is a European-led proposal for an experimental facility to demonstrate the feasibility of laser-driven fusion as a future energy source using the fast-ignition approach. The project is still at the design stage—a preliminary report received positive reviews from the European Community in 2007 but there is still no firm commitment to go ahead.

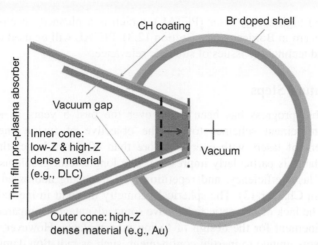

FIGURE 12.9 A schematic view of a new design of the double-cone target. The tip of the cone is extended to guide fast electrons close to the core. The inner cone is made of a relatively dense low-Z material, such as *diamond-like carbon* (DLC), and the outer cone is a dense high-Z material, such as gold.

With the fast-ignition approach, HiPER plans to produce a fusion energy output of the same magnitude as NIF and LMJ but with smaller and less expensive lasers. In the preliminary specification, HiPER's driver would be about 200 kJ and would compress the fuel capsule to a density of about 300 g cm $^{-3}$— about one third of the compressed density reached by NIF with indirect drive—but sufficient for the fast-ignition approach. Ignition would be started by a very short (~10 picoseconds) ultrahigh-power (~70 kJ, 4 PW) laser pulse. Thus, the combined laser energy in HiPER would be about 270 kJ, compared to the 1.8 MJ used in NIF and LMJ. HiPER is expected to be significantly less expensive to build than NIF or LMJ.

HiPER plans to use capsules with gold cones (the technique tested on FIREX) to allow the ultra-fast laser to penetrate into the center of the compressed fuel. HiPER will also study the alternative method, where the laser generates relativistic electrons that penetrate through the compressed fuel. One of the key physics issues for HiPER will be to study the rapid heating processes. To work efficiently, the relativistic electrons have to penetrate into the compressed core and then stop in as short a distance as possible, in order to release their energy into a small spot and thus raise the temperature as high as possible.

In order to make a commercial power plant, the high gain of a device like HiPER will have to be combined with a high-repetition-rate laser and a target chamber capable of extracting the power. HiPER proposes to build a demonstration diode-pumped laser producing 10 kJ at 1 Hz or 1 kJ at 10 Hz (the final design choice is still open). This would be between 10 and 500 times bigger than the best high-repetition-rate lasers currently operating (typically, they are in the range of 20 to 100 J). A forerunner to HiPER will be the multi-petawatt

(10^{15} watts) high-energy laser (PETAL), which is a planned development of the LIL system in Bordeaux (see Section 12.3). PETAL will be used to address physics and technology issues of strategic relevance.

12.7   Future Steps

Considerable progress has been made over the past 5 years in advancing inertial-confinement schemes toward the objective of fusion energy. The development of lasers with energies more than 40 times larger than previously available is particularly impressive. The longer-term problems of target chambers, laser efficiency, and repetition rate are being addressed (and are discussed in Chapter 13). The spherical geometry inherent in inertial confinement and the lack of magnetic fields have some advantages compared to magnetic confinement for the design of fusion power plants. However, there are also problems unique to inertial confinement, such as radiation damage to the optics close to the target chamber by neutrons and X-rays. A number of groups in Japan, Europe, and the US are seriously considering plans for further development of the program for power plants, and collaboration on an international program is a possible way to proceed.

Chapter 13

Fusion Power Plants

13.1 Early Plans

Studies of fusion power plants based on magnetic confinement started as early as 1946, with George Thomson's patent application, as discussed in Chapter 5. A few years later, Andrei Sakharov in the Soviet Union considered deuterium-fueled fusion and Lyman Spitzer in the US discussed prospects for power plants based on stellarators. These initiatives opened the story, but it could go no further because at that time no one knew how to contain or control hot plasma. By the end of the 1960s, there was growing confidence that magnetic-confinement problems would be solved, and this prompted a renewal of activity on the subject of fusion power plants. Bob Mills at Princeton, Fred Ribe at Los Alamos, David Rose at MIT, and Bob Carruthers and colleagues at Culham all looked into what would be involved in building a fusion power plant and identified some of the problems that had to be solved. Fusion power-plant design has been a feature of the magnetic-confinement fusion program ever since. A particularly significant contribution was the ARIES series of studies (mainly magnetic confinement, but there is a study of an inertial-confinement power plant) by a collaboration of the University of Wisconsin, UCLA, UCSD, and other US research institutions.

Having reached the point where research into both magnetic- and inertial-confinement fusion has made remarkable progress in solving many of the physics problems, it is now an appropriate time to consider the many technological and engineering issues involved in building a fusion power plant. Some aspects of fusion power-plant design are common to both magnetic and inertial confinement, but there are some significant differences, which are discussed where relevant. One very important difference (Box 13.1) will be that the hot plasma at the heart of a magnetic fusion power plant will operate in a steady-state or quasi steady-state mode, whereas inertial confinement will be based on repetitively pulsed systems.

There is renewed emphasis on the need to develop fusion energy soon enough for it to have an impact on current options for energy policy. This requires that a demonstration fusion power plant must be brought into operation as early as possible in order to develop and test the technology—paving the way for the construction of the first generation of commercial-scale fusion power plants. To become commercially viable, fusion energy will have to satisfy

Fusion, Second Edition.

Box 13.1 Steady-State versus Pulsed Power Plants

Fusion energy research started out with rapidly-pulsed magnetic-confinement systems, and this led to the famous *Lawson condition* for energy breakeven (see Section 4.2), which was based on the assumption that all of the fusion energy from one pulse would be taken out as heat, converted into electricity, and used to initiate and heat the next pulse. The emphasis in magnetic-confinement fusion is now on quasi-steady-state systems—but inertial-confinement fusion is inherently pulsed. In a repetitively pulsed fusion power plant, some of the energy from one pulse has to be used to drive the next pulse. An important question is: How much energy can be allowed to recirculate? To be commercially viable, a power plant will have to export and sell most of the electricity that it generates and it can afford to use only a relatively small part of its output for the next pulse. This is known as the *recirculating energy fraction, F_r,* and it is equal to the electrical energy that is required to drive the next pulse divided by the total electricity output of the plant. Of course, to some extent all power plants have to use some of their own output to run plant ancillary equipment, such as cooling systems, but with some fusion energy concepts the recirculating fraction could be quite large, and this would have a big impact on the economics of the plant. A reasonable upper limit for the recirculating power fraction in a commercial power plant would be about 20%.

An important figure of merit for magnetic-confinement fusion systems is the *multiplication factor Q,* which is defined as the total output of fusion power (alphas plus neutrons) divided by the input of external heating power needed to maintain a steady-state energy balance in the plasma fuel. (This parameter is usually referred to as *big Q* to distinguish it from *little q*—the parameter known as the *safety factor,* which is discussed in Box 9.1.) We require $Q = 1$ for breakeven and $Q = \infty$ for ignition. ITER is designed to reach $Q \approx 10$ (with fusion power of about 500 MW and external heating power of 50 MW) and this is generally taken as being the basic requirement for achieving a *burning plasma*—where the alpha particle heating (20% of the fusion output power) starts to dominate the energy balance. At $Q = 10$, the internal alpha particle heating (100 MW) is twice as large as the external heating. In magnetic confinement, ignition is the point where the alpha particle heating takes over completely and the external heating can be switched off. We generally assume that a magnetic-confinement fusion power plant would run close to this self-sustaining ignited state. In fact, some recirculating power will be required to provide additional heating for fine control of the plasma and external drive to maintain a non-inductive plasma current, and if this is taken into account, a typical value would be $Q \approx 20$.

Inertial-confinement fusion is inherently repetitively pulsed, and an important parameter is the target gain G, which is equal to the fusion energy output divided by the driver energy into the target chamber. It is important to remember that only a small fraction of the driver energy that enters the target chamber ends up as heat energy in the capsule—some of the driver energy is lost in the conversion into X-rays in the hohlraum and most of the X-ray energy goes into heating and accelerating the ablator. For ICF, the parameter Q (sometimes referred to as the *physics Q*) is defined as the fusion energy output divided by the actual external heat energy input into the plasma—not the driver energy. Ignition, in ICF terms, is the condition where the core of the compressed fuel capsule becomes sufficiently hot and dense that the localized fusion heating (by alpha particles) takes over and a *burn wave* propagates outward through the surrounding mass of compressed fuel. ICF researchers sometimes refer also to a quantity known as *engineering* Q_E, which is defined as the electrical power output divided by the electrical input—therefore we have $Q_E = 1/F_r$, and $Q_E = 5$ corresponds to 20% recirculating energy.

many stringent (and sometimes conflicting) criteria, including the capital cost and timescale of building the plant, the cost of electricity, environmental impact, safety, and public acceptability. Particularly important issues are reliability, availability, and ease of maintenance. Unplanned and unpredictable shutdowns of large-scale power plants have a serious impact on the stability of the electricity grid. Utility companies need to be able to predict maintenance shutdowns in advance and to schedule them for periods of low demand for electricity. Fusion faces a very difficult challenge as a fledgling (and very advanced) technology in meeting the high levels of availability and reliability that have been established as the norm by other energy technologies.

13.2 Fusion Power-Plant Geometry

A fusion power plant will be built up in a series of concentric layers—like the layers in an onion—as shown schematically in Figure 13.1. The onion concept is particularly appropriate for an inertial-confinement power plant, where the topology will be spherical, but the layers in a magnetic-confinement power plant based on the tokamak will be toroidal and D-shaped in cross-section. The *burning plasma* forms the core, and the material surface that surrounds it is known as the *first wall*. Outside the first wall will be the *tritium-breeding blanket*, followed by a *neutron shield*, the *vacuum vessel*, the *magnetic coils* (in the case of magnetic confinement), and, finally, a second shield to reduce the radiation down to the very low level required for personnel working nearby.

The average power flux at the first wall is typically several megawatts per square meter (MW m^{-2}), but it is important to remember that 80% of the power from the DT reaction is carried by neutrons, which pass through the wall without depositing much heat in it. The neutrons slow down in the blanket region, depositing most of their energy as heat within the first half meter or so (the exact depth of the deposition depends on the composition of the breeding and structural materials) before they finally react with lithium to produce tritium. The other 20% of the DT fusion energy is released as alpha particles—this energy goes into heating the plasma and ultimately ends up leaving

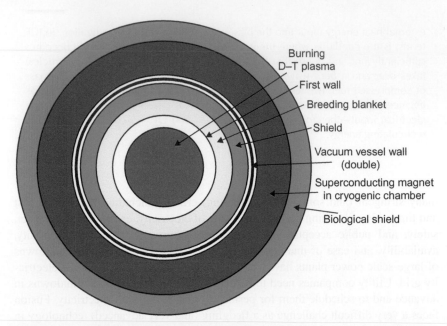

Burning
D–T plasma

First wall

Breeding blanket

Shield

Vacuum vessel wall
(double)

Superconducting magnet
in cryogenic chamber

Biological shield

FIGURE 13.1 Cross-section of a conceptual design for a fusion power plant. The power is extracted from the blanket and is used to drive a turbine and generator, as shown in Figure 1.2. For illustration, the cross-section has been shown circular, which would be the case for inertial confinement, but in a tokamak it would probably be D-shaped. No magnets are required for inertial confinement.

the plasma edge as ions and electrons and as ultraviolet, visible, and infrared radiation. In an inertial-confinement power plant, all of this energy from the plasma will be deposited on the first wall in the form of radiation and energetic particles. In a magnetic-confinement power plant, a significant fraction of the plasma energy will be "siphoned off" by the divertor. If deposited uniformly, the typical direct heat flux onto the first wall will be less than 1 megawatt per square meter. The first wall, the blanket, and the divertor will get hot and require cooling by high-pressure water or helium gas, so that the fusion energy can be taken out and converted into electricity. The primary coolant will pass through a heat exchanger and produce steam, which will be used to drive turbines and generators, as in any conventional power plant (Figure 1.2). Economical operation of a fusion power plant will require high power fluxes and high thermal efficiencies, thus demanding high operating temperatures. Structural materials tend to deform when operated under these conditions, and this will be a controlling factor limiting the choice of materials.

A high power flux puts severe constraints on the mechanical and thermal design of the plasma-facing surfaces of the first wall. The problem is complicated because the surfaces facing the plasma are eroded by particles and radiation. The choice of material for these surfaces is therefore constrained by the necessity to use materials that minimize erosion and have good heat resistance. For magnetic confinement, the problem is most severe at the divertor target plates, which are subject to highly localized heat fluxes. For inertial confinement, the pulsed operation poses special problems for the first wall of the target chamber.

The toroidal construction of a magnetic-confinement power plant traps the inner layers within the outer layers, so maintenance and repair will be difficult, and this must be taken into account at the design stage. The spherical construction makes an inertial-confinement power plant simpler to maintain, at least in principle. In both cases, the blanket and first wall will have penetrations for heating and exhaust systems and for diagnostics. Some components will require regular replacement during the operating life of the plant. In both magnetic and inertial confinement, the structure surrounding the plasma will become radioactive, and "hands-on" work will not be possible after the plant has been put into operation; thus, remote-handling robots will have to carry out all of this work. Some of the techniques to do this are being developed and tested in present-day experiments, like JET and ITER.

13.3 Radiation Damage and Shielding

Fusion neutrons interact with the atoms of the walls, blanket, and other structures surrounding the plasma. They undergo nuclear reactions and scatter from atoms, transferring energy and momentum to them. Neutrons have a number of deleterious effects. First, they damage the structural materials. Second, they cause the structure to become radioactive, which requires the material to be carefully recycled or disposed of as waste at the end of the power plant's life. However, the activity of materials in fusion power plants is confined to the structural materials, since the waste product of the fusion reaction is helium.

Radiation damage processes (Box 13.2) have been studied in considerable detail in fission reactors, providing a good basis for assessing the problems in fusion power plants. The fusion neutron spectrum consists of a primary component of 14 MeV neutrons from the DT reaction together with a spectrum of lower-energy neutrons resulting from scattering. The damage caused by fusion neutrons is expected to be more severe than that from fission neutrons because the spectrum of fusion neutrons extends to higher energies. The higher-energy neutrons cause reactions that deposit helium in the solid lattice, and this has a marked effect on the behavior of materials under irradiation. To fully characterize the materials for a fusion power plant requires the construction of a powerful test facility with neutrons in the appropriate energy range.

Box 13.2 Radiation Damage

When an energetic neutron collides with an atom of the first wall or blanket structure, it can knock the atom out of its normal position in the lattice. The displaced atom may come to rest at an interstitial position in the lattice, leaving a vacancy at its original lattice site (Figure 13.2). In fact, the displaced atom may have enough energy to displace other atoms before coming to rest, and so the damage usually occurs in cascades. The damage is quantified in terms of the average number of displacements per atom (dpa) experienced during the working life of the material. In places like the first wall, where the neutron flux is highest, the damage rate is expected to reach the order of hundreds of dpa. At these levels of damage, the strength of the material will be reduced significantly, and some of the wall components will have to be renewed several times during the lifetime of the power plant. In principle, the neutron damage can be reduced by reducing the neutron flux—but this requires a larger structure for the same output power and increases the capital costs. Thus, there has to be a careful optimization of the size of the plant to strike a balance between the capital and maintenance costs.

A neutron can also undergo a nuclear reaction with a lattice atom, leaving a transmuted atom (or atoms) in place of the original. Usually the new atom is radioactive, and this is the main source of radioactivity in a fusion power plant. The neutron can also eject a proton or an alpha particle from the target nucleus, and these reactions are referred to as (n,p) and (n,α) reactions. A typical (n,p) reaction is

$$^{56}\text{Fe} + \text{n} \rightarrow \text{p} + {}^{56}\text{Mn}$$

The (n,α) reactions are particularly important for the 14 MeV neutrons from the DT fusion reaction. The protons and alpha particles produced in these reactions pick up electrons and form hydrogen and helium atoms in the lattice. Individual atoms tend to coalesce, forming bubbles of gas in the lattice, and can be deleterious to the structural strength. Further damage to the lattice can be produced by energetic recoil of the reaction product.

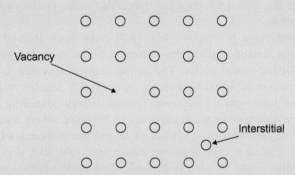

FIGURE 13.2   Schematic of the formation of interstitials and vacancies in a lattice structure. Hydrogen and helium atoms are also introduced into the lattice due to (n,p) and (n,α) reactions.

13.4 Low-Activation Materials

One of the most important advantages of fusion relative to fission is that there will be no highly radioactive waste from the fuel cycle. The structure will become radioactive by exposure to neutrons, but careful choice of the materials means that the radioactivity of a fusion power plant will fall to a very low value in less than 100 years, as shown in Figure 13.3—this is also discussed in Chapter 14. The blanket, the first wall, and (in the case of magnetic confinement) the divertor systems will have a finite operational life and must be maintained and periodically replaced using robotic systems. Initial levels of radioactivity of material removed from service and the rate of decay of the various radioactive isotopes will dictate the acceptable storage and disposal methods and the possibility of recycling used components. Recycling materials, which would reduce the amount of radioactive waste, may be a possibility. To achieve low activation, it is necessary to choose the structural materials carefully. An important source of activation comes from small traces of impurities in the structural materials, rather than from their main constituents, and careful control of quality and material purity will be critical to achieving the low-activation goal.

Although the successful development of suitable low-activation structural materials is a formidable challenge, they would be a significant improvement over presently available materials like austenitic stainless steel. Three groups of materials are under consideration as low-activation structural materials (Box 13.3). At present the low-activation martensitic steels are best understood, and they are the most probable candidates for a fusion power plant in the medium term. However, development of suitable silicon carbide composites would have the advantages of higher power-plant operating temperatures,

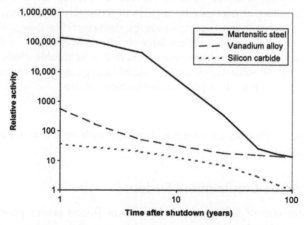

FIGURE 13.3 Calculated activity of a low-activation martensitic ferritic steel, vanadium alloy, and silicon carbide as a function of time from shutdown of power-plant operation.

Box 13.3 Low-Activation Materials

There are three groups of materials with potential as low-activation structural materials. The first group is known as *martensitic steels*. These are low-carbon steels based on iron, with typically 12% chromium and 0.12% carbon. They are used extensively in high-temperature applications throughout the world, including fission reactors. The database on their properties in a radiation environment is extensive. The upper operational temperature limit, based on high-temperature strength properties, is usually estimated to be 550°C. The main constituents of martensitic steels, iron and chromium, have relatively good activation properties, and the main source of activation comes from some of the minor ingredients, like niobium and molybdenum. Low-activation martensitic steels are being developed by reformulating the composition (for example, replacing niobium with vanadium or titanium and replacing molybdenum with tungsten). These reformulated steels have properties very similar to (and in some cases superior to) those of the existing alloys. Further studies are needed on the effects of helium deposition in the lattice.

Alloys of *vanadium* form the second group of materials, and the radiation-induced characteristics are being studied specifically for the fusion materials program. The elements vanadium, titanium, chromium, and silicon, which make up most vanadium alloys, all qualify as low activation. Some vanadium alloys have been studied in the US, but it has been found that they become brittle below 400°C and above 650°C, leaving a rather narrow operating range.

The third material under study is *silicon carbide* (SiC), a material that is being developed for aerospace applications. Interest in the use of silicon carbide composites stems not only from the low-activation properties but also from the mechanical strength at very high temperatures (1000°C). This might make possible a high-temperature direct-cycle helium-cooled power plant. However, under irradiation there are indications that swelling of the silicon fiber and decreased strength can occur at temperatures in the range of 800 to 900°C. In addition to developing composites that have adequate structural strength, there are other technical problems to be solved. Methods must be developed that will seal the material and reduce its permeability to gases to acceptably low figures. Methods of joining SiC to itself and to metals must be developed. The design methodology for use of SiC composites in very large structures, at high temperatures with complex loads, and in a radiation environment is a formidable challenge. It is recognized that development of suitable materials is going to be of comparable difficulty to the solution of the plasma confinement problem.

and hence higher thermal efficiencies, together with very low activation rates for maintenance.

13.5   Magnetic-Confinement Fusion

The minimum size of a magnetic-confinement fusion power plant is set by the requirement that the plasma has to be big enough to achieve a sufficiently large value of Q (see Chapters 4 and 10 and Box 13.1). On the basis of the ITER scaling for energy confinement (Box 10.4), the minimum size for a tokamak power plant that would reach ignition ($Q = \infty$) would have physical dimensions slightly larger than the present ITER design (Table 11.1) and the power output would be about 3 GW (gigawatts) of thermal power, which would correspond to about 1 GW when converted into electricity. However, there are many other factors that will determine the optimum size and the value of Q for a commercial power plant.

An important difference between magnetic and inertial confinement is the magnetic field. Most magnetic-confinement experiments use copper coils cooled by water. When scaled up to ITER size, copper coils would consume a large fraction of the electricity generated by the power plant, and the net output of the plant would be reduced. The solution is to use superconducting coils—made from special materials that, when cooled to very low temperatures, offer no resistance to an electric current. Once energized, a superconducting magnet can run continuously, and electricity is required only to run the refrigerator that keeps the coil at the low temperature. Some present-day tokamaks have been constructed using superconducting coils, including TRIAM, a small Japanese tokamak that has maintained a plasma for many hours; Tore Supra at Cadarache in France; and tokamaks in Russia, China, and South Korea. There is also a large superconducting stellarator operating in Japan and one being built in Germany. Present-day superconducting materials have to be cooled to very low temperatures using liquid helium, but recent developments in superconducting technology may allow magnets to be made out of materials operating at higher temperatures, which would simplify the construction. Superconducting coils have to be shielded from the neutrons (Box 13.4) and therefore will be located behind the blanket and neutron shield. A radial thickness of blanket and shield of about 2 meters is needed to reduce the neutron flux to a satisfactory level. The neutron flux is attenuated by many orders of magnitude in passing through the blanket and neutron shield, as shown in Figure 13.4.

Some of the fusion power transferred to the plasma from the alpha particles is radiated and deposited fairly uniformly on the first wall. If all the loss from the plasma could be transferred uniformly to the wall, the flux (about $0.5 \, MW \, m^{-2}$) would be relatively easy to handle, but there are various reasons why it is difficult to radiate all the power and why some power has to be removed via the divertor. Localized power loads on the divertor plate are inherently high (Box 13.5), and this is one of the most critical elements in designing a tokamak power plant. The divertor also removes the alpha particles after they diffuse out of the plasma and become neutral helium gas. This gas has to be pumped out to prevent it from building up in concentration and diluting the DT fuel. A maximum concentration of about 10% helium in the plasma is normally considered to be the upper limit.

Box 13.4 Shielding the Superconducting Coils

The superconducting coils have to be placed behind the blanket and neutron shield in order to protect them from the neutrons. Some of the neutron energy is deposited as heat in the superconducting magnets and this must be kept within stringent limits. The liquid helium refrigerators needed to keep the magnets at very low temperature consume typically 500 times the energy they remove from the low-temperature coil. The refrigerators have to be powered from the power plant output, and obviously this forms part of the recirculating energy and needs to be kept as small as possible. Taking into account the efficiency of converting fusion energy as heat into electricity (typically 33%), in order to keep the consumption of power by the refrigerators to less than 3% of the plant output, it is necessary to shield the superconducting coils so that they receive less than 2×10^{-5} of the neutron power flux at the first wall. This criterion will be less strict if high-temperature superconductors can be developed for large high-field coils.

A further factor to be considered is the radiation damage in the component materials of the coils. The most important are the superconductors themselves, normal conductors such as copper (which are included to stabilize the superconductor and to protect it under fault conditions), and the insulators. Superconducting materials have limited ability to withstand radiation damage. Irradiation with high doses leads to a reduction in the critical current, above which the material reverts from a superconducting state to a normal resistive material. An upper limit of the total tolerable neutron dose is about 10^{22} neutrons m^{-2}, corresponding over a 20-year lifetime to an average flux of about 2×10^{13} neutrons $m^{-2}s^{-1}$. Although further work is needed to assess damage in normal conductors and insulators, the indications are that these materials should be satisfactory at the fluxes that are determined by the criteria for the superconductor itself.

13.6   Conceptual Power-Plant Studies and DEMO

The European Power Plant Conceptual Study (PPCS) was carried out between 2001 and 2005 to compare five conceptual designs for a magnetic-confinement fusion power plant based on the tokamak. All of the conceptual designs have a net electrical output of about 1500 MW(e) and they cover a wide range of extrapolations in technology and physics. The power-plant concept (model A) that is based on technology and physics with the smallest extrapolation from ITER would have the largest physical dimensions (major radius $R \approx 9.5$ m and plasma current $I \approx 30$ MA). A design concept (model D) that would be close to ITER in physical dimensions and current ($R \approx 6$ m, $I \approx 14$ MA) would require substantial advances in technology and tokamak physics. In between these two extremes there are conceptual designs that rely on progressive advances of tokamak physics and fusion technology together with the development and validation of low-activation structural materials (Box 13.3).

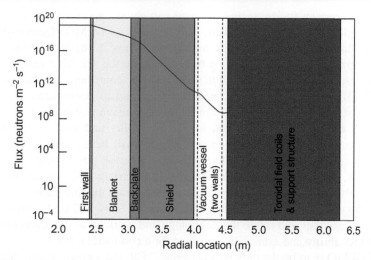

FIGURE 13.4   Calculated neutron flux as a function of the minor radius in a typical magnetic-confinement fusion power plant with neutron power flux of 2.2 MW m^{-2} at the first wall. The locations of the main components are indicated. The design is water cooled, with a lithium–lead breeder blanket.

Box 13.5 Power Handling in the Divertor

The magnetic geometry constrains the plasma to follow field lines and deposits the power on the target plate in the divertor (Figure. 9.5). The width of the zone in which the power is deposited on the divertor target is determined by the magnetic geometry and by the inherent cross-field diffusion of the plasma—both factors tend to deposit the power in a relatively narrow annular zone. One design solution to increase the area exposed to plasma is to incline the divertor target at a steep angle relative to the magnetic field. A second solution is to reduce the power flux to the target by increasing the fraction that is radiated in the plasma edge. But to radiate a large amount of power, it is necessary to introduce impurities. There is a limit to how much impurity can be introduced to cool the edge without cooling the center of the plasma. There are also limits to the fraction of power that can be radiated without driving the plasma unstable.

Maintenance is a major factor in determining the availability of a fusion power plant and it is expected that erosion of the divertor, where much of the plasma energy is deposited, will require it to be replaced roughly every 2 years. The mechanical properties of the blanket will be degraded by neutron damage and it may have to be replaced every 5 years. An availability of around 75–80% should be possible for a first-generation fusion power plant. The development of new materials and operating procedures to reduce the need for maintenance and to extend the plant's availability are important areas of future development.

The European fusion energy program has always assumed that there will be a single intermediate step, known generically as DEMO (Demonstration Power Plant), to go from ITER and IFMIF (Box 13.6) to a commercial fusion power plant. ITER's goal is to demonstrate the scientific feasibility of magnetic-confinement fusion and to make a strong contribution to the technology. IFMIF's task is to develop new materials and to demonstrate that they can withstand the environmental conditions in a fusion power plant. DEMO's role will be to confirm the technological feasibility and to demonstrate the commercial viability of fusion energy. DEMO will be the first fusion plant to generate electricity and it is likely to have a fusion power output of about 1500 MW, corresponding to about 500 MW of electricity. It is usually assumed that DEMO will have an initial phase, to demonstrate the integration of fusion technology, materials, and physics developments. This will be followed by a second phase that will refine the technology and work up to the levels of operational reliability and availability required for a commercial plant.

If DEMO is to be the only step between ITER and a power plant, there are many difficult questions to be addressed before its design can be finalized. An important consideration is to decide where exactly DEMO should lie in the range of development between ITER and a commercial power plant. If DEMO is built on the basis of technology that is relatively close to that of ITER, the next step to a power plant will be large and might require an intermediate stage. On the other hand, DEMO's design and construction will be delayed if it is based on very advanced technology. This question is pertinent following

Box 13.6 The Broader Approach

During the ITER site negotiations, the European Union and Japan agreed to fund a Broader Approach to the international fusion program. The Broader Approach will concentrate on areas requiring further work for the development of a fusion power plant. Emphasis will be put on the development of new materials, and a Fusion Materials Irradiation Facility (IFMIF) will be designed. Further materials research is also required to optimize existing low-activation materials in order to minimize the residual activity of a fusion power plant. The development of more advanced materials, such as silicon carbide composites, is also highly desirable; this would enable operation of the plant at higher temperatures and hence with higher thermodynamic efficiencies. The cladding on the walls that face the plasma, and that protects the structural components from the heat from the plasma, also requires further study and development.

A Fusion Energy Research Center will have three divisions, one for power plant design and research coordination, one for computational simulation, and one for the remote participation in the ITER experiment. In addition, an advanced superconducting tokamak based on the existing JT-60U tokamak at Naka will be constructed (Figure 10.10). It will be used to try out operating scenarios, diagnostics, and materials prior to incorporating them in ITER.

a recent European strategic review of energy technology that considered whether fusion energy development can be accelerated to meet the growing demands for carbon-free sources of energy. One option might be to start construction of DEMO earlier than had been assumed—perhaps even before the ITER and IFMIF programs have yielded their full results. In this context, one of the many difficult questions facing DEMO is whether it should be designed to operate steady state or pulsed.

The current in a tokamak like ITER is driven *inductively* and its duration is limited by the physical size of the central solenoid. The inductive current in ITER can be maintained for about 8 minutes, but a pulsed DEMO would require a pulse length of about 8 hours. The central solenoid will have to be correspondingly larger, resulting in a machine with major radius of more than 9 m. The pulse duration of a tokamak can be extended indefinitely if the inductive current is replaced by *non-inductive* current, and there are two mechanisms that allow this. The first mechanism occurs naturally as a result of the trajectories of the plasma electrons in the tokamak magnetic fields. This self-generated non-inductive current is known as the *bootstrap current*. It has been verified experimentally and is well understood in terms of neoclassical theory (Box 10.3). The bootstrap current will contribute between 45% and 75% of the total current in a fusion power plant. The second mechanism to drive a non-inductive current uses the neutral beam and radiofrequency plasma heating systems to deliver energy to the plasma electrons, preferentially in one direction around the torus. Both non-inductive current drive mechanisms need to be optimized and they need to be used together in order for a fusion power plant to operate in steady state. This stage of development will be relatively late in ITER's research program. Thus, the question now being asked is whether DEMO should go ahead earlier as a pulsed design without waiting for ITER to demonstrate reliable steady-state operation with a fully non-inductive current.

Groups in many countries are working on designs for DEMO. The option is still open for individual countries or groups of countries to build such a project as part of a national or regional program. If the administrative structure for ITER works well, it is possible that DEMO may also be built by a large international consortium. A design for DEMO is one of the aims of the collaborative program between Europe and Japan known as *The Broader Approach* (Box 13.6).

Conceptual studies of fusion power plants like the European study discussed earlier tend to focus on a unit size of about 1500 MW electrical output, which is generally considered the maximum that can be integrated into a national electricity grid. The economics of fusion power tends to improve with increasing power output from the plant, and there is growing interest in the concept of a larger fusion power plant that would manufacture hydrogen as well as generating electricity. Hydrogen is a carbon-free fuel that can be used to replace petroleum in transport and other applications. It can be produced from water by electrolysis and by other thermo-chemical cycles.

13.7   Inertial-Confinement Fusion

Some of the component parts of an inertial-confinement fusion power plant would be broadly similar to the equivalent parts of a magnetic fusion power plant—for example, the blanket, heat transfer systems, neutron shielding, and electricity generators. The major differences lie in the target factory, the laser drivers, and the target chamber itself (Figure 13.5). One of the potential advantages claimed for inertial confinement is that some of the high-technology equipment—especially the target factory and the driver—could be located some distance away from the fusion chamber, leading to ease of maintenance. However, the target-handling systems and some optical components (including windows) have to be mounted on the target chamber, where they will be subject to blast and radiation damage. The pulsed nature of inertial-confinement fusion poses special problems for thermal and mechanical stresses in the first wall of the target chamber. Each capsule explodes with energy equivalent to several kilograms of high explosive—the first wall will have to withstand the blast waves from these explosions, and the chamber and the blanket structures will have to withstand the repetitive thermal and mechanical stresses. Such pulsed operation can lead to metal fatigue and additional stress above that of a continuously operating system. Some

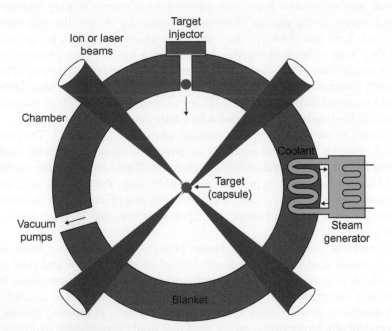

FIGURE 13.5 Proposed design for an inertial-confinement power plant. A target capsule is injected into an evacuated chamber, and a pulse of laser light is fired at it to compress and heat it. The conceptual design is similar to that shown in Figure 13.1, but there are no magnet coils. Heat extracted from the blanket will go to a conventional steam plant.

fusion-chamber concepts incorporate thick liquid layers or flowing layers of granules on the first wall to protect it.

Present-day inertial-confinement experiments shoot less than one target per hour—in some cases only one per day. An inertial-confinement power plant will require about 10 shots every second—over 100 million each year. A big challenge is to develop driver systems with the necessary efficiency, reliability, and repetition rate. The lasers that lead the way in present-day experiments, such as NIF, convert less than 1% of their electric input into target drive energy and they are much too inefficient to meet the demanding requirements of a power plant. The LIFE study (which is discussed in Section 13.8) is optimistic that these problems can be solved with ongoing developments in solid-state laser technology. Other driver technologies, such as ion accelerators and intense X-ray sources, might be developed as alternatives, and new concepts, such as fast ignition and shock ignition (Boxes 7.5 and 12.3), might go some way to reduce the requirements. Target design and fabrication will need substantial development. The targets for present-day experiments are manufactured individually and assembled by hand using fabrication techniques that are not well suited to economical mass production. To generate electricity at a competitive price, the cost of a target will have to fall substantially. This will require a highly automated target factory with a substantial development in technology.

13.8   A Demonstration ICF Power Plant—LIFE

With NIF now operating successfully and expected to achieve ignition for the first time by the end of 2012, thoughts at Livermore have turned to designing a demonstration inertial-confinement fusion power plant based on the physics that is being tested in NIF. A fundamental issue for an ICF power plant (as discussed in Box 13.7) is to have a driver that can convert electrical energy with sufficiently high efficiency into the laser energy that is required to compress and heat the fusion target, and to combine this with a target that has sufficiently high energy gain that there will be a substantial net output of energy. The LIFE design study is based on a laser driver delivering 2.2 MJ of energy at a wavelength of 351 nm with efficiency of 18% (a net value of 15% when the energy required to cool the laser is taken into account) and a target with gain in the range of 60 to 70. This is a substantial extrapolation from NIF, where the driver efficiency is about 0.5% and the target gain is expected to be in the range of 10 to 20. The LIFE study has carried out detailed calculations of the nuclear reactions in the tritium-breeding blanket and finds that the blanket could produce a gain in energy by a factor 1.2 and have a tritium-breeding ratio of 1.1. It is planned to run the blanket and first wall at a relatively high temperature in order to allow the generator to run on a supercritical steam cycle with a heat-to-electricity conversion efficiency of 44%. As discussed in detail in Box 13.7 and illustrated in Figure 13.6, these conditions are consistent with about 20% of the output energy being used by the laser driver, allowing the remaining 80% to

Box 13.7 Driver Efficiency and Target Gain

Inertial-confinement fusion is inherently a repetitively pulsed process (see Box 13.1). The laser driver has to provide sufficient energy to compress and heat the capsule containing the fusion fuel to the point that it ignites and burns—and some of the fusion energy that is released has to be used to drive the next pulse. To be commercially viable, an ICF power plant can afford to use only a relatively small part, let us say 20%, of its output to provide for the driver. It is important to consider the minimum values of the target gain and the driver efficiency that will be required for an ICF power plant.

The *target gain G* is defined as the fusion energy released when the capsule burns divided by the driver energy E_D that is delivered by the laser into the target chamber. (As we discuss below, only a small fraction of the driver energy that enters the target chamber ends up as heat in the capsule—some of the driver energy is lost in the conversion into X-rays in the hohlraum and most of the X-ray energy goes into heating and accelerating the ablator.) The fusion energy released by each capsule will be GE_D, and this is deposited as heat in the first wall and the blanket of the target chamber. The heat energy has to be taken out and converted into electricity before it can be used by the driver, and the conversion efficiency from heat to electricity is constrained by the laws of thermodynamics. If the heat-to-electricity conversion efficiency of a typical power plant were 33%, the total electrical output would be $0.33GE_D$. If we assume that the driver uses 20% of the total electricity output and ε_D is the efficiency of the driver at converting this electricity into driver energy delivered into the target chamber (ε_D is sometimes referred to as the *wall-plug efficiency*), we have $E_D = \varepsilon_D(0.2 \times 0.33)GE_D$, and so $\varepsilon_D G = 15$.

The LIFE design study assumes that there would be an enhancement of the heat energy in the blanket by a factor 1.2 due to exothermic tritium-breeding reactions and that heat-to-electricity conversion efficiency as high as 44% could be obtained by using a super-critical steam cycle at wall temperatures up to 600°C. LIFE requires a target gain G in the range 60 to 70 with driver efficiency $\varepsilon_D = 18\%$ (electricity to laser energy), which comes down to 15% when the energy needed to cool the laser is taken into account. The range of values of target gain and driver efficiency required for an ICF power plant within these assumptions is plotted in Figure 13.6. The calculations for NIF have predicted a target gain in the range $10 < G < 20$ for the baseline indirect-drive target. The NIF laser uses about 350MJ of electrical energy to produce the 1.8MJ of ultraviolet light that enters the target chamber, so this corresponds to a driver efficiency $\varepsilon_D = 1.8/350 \approx 0.5\%$. The step from NIF to LIFE will require substantial improvements in target gain and in driver efficiency, as discussed in the main text.

The minimum driver energy required to reach ignition is a particularly important parameter and it is convenient to write it as $E_D = E_C / \varepsilon_C$, in terms of the minimum energy E_C that has to be absorbed by the capsule to compress and heat it sufficiently to ignite, and the overall efficiency ε_C with which the laser energy is transferred into the capsule. The transfer efficiency ε_C takes into account the conversion of laser light into X-rays in the hohlraum and the transfer of X-ray energy into the capsule—it can be optimized by the choice of the geometry and materials of the hohlraum and by the material used for the capsule ablator, but clearly we will have $\varepsilon_C \ll 1$ and therefore $E_D \gg E_C$. Calculating these quantities requires sophisticated physics models and state-of-the-art computations that are well beyond the scope of our discussion here, but it turns out that the results of these calculations can be fitted empirically by $E_C \sim \alpha^{1.9}/(u^{5.9}P^{0.8})$, which allows us to have some insight into the important parameters. According to this empirical formula, the absorbed capsule energy (and therefore the required driver energy) depends very strongly on u (the *implosion velocity* of the capsule) and also on P (the *ablative pressure* driving the implosion). So it is very desirable to optimize these quantities—especially the implosion velocity, where a relatively small increase in velocity (say 5%) would result in a large reduction in the required driver energy (perhaps by as much as 30% according to the empirical formula)—but the implosion becomes unstable if either the velocity or the pressure is too large, so this is a delicate balance.

The parameter α is called the *in-flight isentrope* and it is a measure of how much more difficult it is to compress the capsule compared to an ideal gas—so it is important to keep $\alpha \approx 1$, and this requires careful control of the energy profile of the laser pulse to avoid preheating the capsule. The calculations for the NIF indirect-drive target assume that $\alpha = 1$ and predict $P = 120$ Mbar and $u = 3.8 \times 10^7$ cm s^{-1}. The driver energy specified for NIF is 1.8 MJ, which includes a safety margin over the minimum values predicted by the calculations—so if everything works perfectly and all the many parameters can be optimized at the same time, it might be possible to reduce the driver energy requirements.

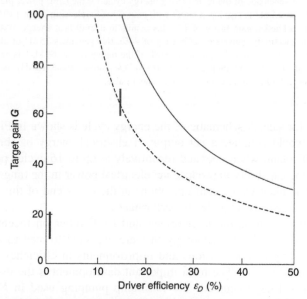

FIGURE 13.6 The fusion gain G required for an ICF power plant plotted against the driver efficiency ε_D assuming that the recirculating energy fraction is 20% (i.e., 80% of the generated energy is exported and 20% is used by the laser driver). As discussed in Box 13.7, the upper (solid) curve assumes that the conversion efficiency from heat to electrical energy is 33%. The lower (dashed) curve assumes that the conversion efficiency is 44% and there is enhancement of the heat energy in the blanket by a factor of 1.2 (as in the LIFE design study). The red vertical bar indicates the values of $60 < G < 70$ and $\varepsilon_D = 15\%$ that are proposed by the LIFE design and the blue vertical bar indicates the values ($10 < G < 20$ and $\varepsilon_D \approx 0.5\%$) expected in the NIF experiment.

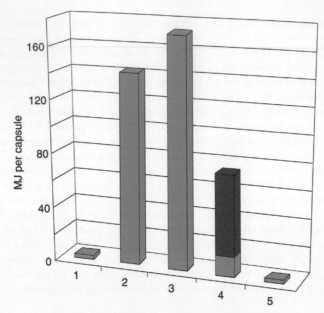

FIGURE 13.7 A schematic of the recirculating energy cycle for the LIFE power plant. The laser driver energy $E_D = 2.2\,MJ$ (1) ignites a target with gain $G \approx 65$, releasing about 140 MJ of fusion energy (2). With a blanket gain factor of 1.2, this results in overall heat energy of about 170 MJ (3). With heat-to-electricity conversion efficiency of 44%, this generates 75 MJ of electricity (4). The upper part of (4), corresponding to about 80% of the output (about 60 MJ), is exported—and the lower part of (4), corresponding to about 20% of the output (about 15 MJ), is used by the driver to produce 2.2 MJ (5) of laser energy for the next pulse.

be exported for sale. A schematic of the energy cycle is shown in Figure 13.7. Each cycle would generate a net output of electrical energy of about 60 MJ, and the LIFE plant, when operated repetitively at up to 16 cycles per second, would have the capability to produce net electrical power in the range of 400 to 1000 MW. The plant would start operations at the lower end of this range and would be progressively upgraded in performance.

A key component is the laser driver, and LIFE relies on recent developments in solid-state laser technology that are expected to lead to significant reductions in physical dimensions and improvements in efficiency compared to the laser used in NIF. The most important development is the use of diode pumping (Box 13.8) to replace the flashlamp pumping used in NIF. Diode pumping would be much more energy efficient, have much higher reliability, and would require much less maintenance. The cost of the diodes has a major impact on the cost of the laser and diodes at present-day prices would be too expensive to use on the required scale—but it is anticipated that costs will fall to an acceptable level with substantial volume production. The designs that are being developed for LIFE are based on the concept of a large number of self-contained laser modules using neodymium-glass amplifiers, diode pumping,

Box 13.8 Diode-Pumped Lasers

The lasers used in present-day ICF experiments are based on slabs of neodymium-doped glass that are pumped by *flashlamps* (also known as *flashtubes*). These are long glass or quartz tubes with electrodes at either end that are filled with a gas, such as xenon or krypton. When a high-voltage pulse of electricity is applied to the electrodes, the gas ionizes and produces an intense flash of light. Small flashlamps are familiar to us as the built-in flash units in modern digital cameras, where they are used to emit a bright flash of white light to illuminate a subject. Flashlamps are used to pump lasers because of their high intensity and brightness at short wavelengths and because of their short and controllable pulse durations. Some of the light emitted by the flashlamp is absorbed by the neodymium atoms, exciting these atoms into the higher energy state that is subsequently discharged as the laser fires (see Box 7.3)—but the process is very inefficient because flashlamps emit light over a broad spectrum and only a small fraction of this energy is absorbed by the neodymium atoms. A more efficient approach would be to pump the laser using light that carefully matches the specific absorption lines of the neodymium atoms, and this can be achieved if laser diodes are used in place of flashlamps.

A diode is an electronic device through which electricity can pass in only one direction. Solid-state diodes are made from a semiconductor material like silicon or gallium arsenide that is treated so that in one half of the diode the electric current is carried by negatively charged electrons and in the other half by positively charged "holes." The electrons and holes combine at the junction between the two halves of the diode to cancel out their opposite electric charges—and in an ordinary diode made of silicon (widely used in many electronic devices) this just maintains the flow of electric current. But in some semiconductor materials, like gallium arsenide, there is an energy gap at the junction and this energy is released as a photon of light when the electrons and holes combine—and because the energy gap has a well-defined value, the light that is emitted has a well-defined wavelength (or color). This basic process can be used to make light-emitting diodes and laser diodes that are very efficient at converting electrical energy into light energy and the wavelength (or color) of the light can be determined by selecting the appropriate semiconductor material and by other factors that determine the energy gap. These devices are widely used in domestic light sources, CD and DVD players, barcode readers, laser printers, and many other applications.

Diode-pumped solid-state lasers are replacing flashlamp-pumped lasers in many scientific and industrial applications. The overall efficiency of a diode-pumped solid-state laser can be as high as 10–20%, due to the high efficiency of the diodes in converting electrical energy into light and also due to the efficiency with which the light can be coupled into the neodymium atoms. However, using diode-pump sources at the megajoule-scale of inertial-confinement fusion lasers requires very high peak power with large arrays of diodes, rather than the single diodes that are used in smaller lasers. The cost of the diodes becomes an important factor—but prices are expected to fall with ongoing developments and especially from large-scale production. Another crucial advantage will be that laser diodes will be able to fire up to ten billion times without being replaced, whereas the large flashlamps used in present-day ICF lasers have to be replaced after about 100,000 shots—and in an application like LIFE, this would be less than a single day's operation.

and helium cooling. The modular concept, combined with sufficient redundancy in the numbers of modules and other components, would allow individual laser modules to be removed and replaced for repair and maintenance even while the power plant remains in operation. This is an essential step toward achieving the high levels of reliability and availability required for a commercial fusion power plant.

The high repetition rate required for LIFE will have an important effect on many components of the power plant, both in terms of the repetition rate itself and the number of shots integrated over a period of time. Thus, the laser—which in NIF fires typically one shot per day—will have to be capable of firing up to 16 shots per second and operating reliably for more than one million shots per day. There is a similar impact on target production, injection, and alignment. The ICF program to date has developed techniques to produce a wide range of sophisticated experimental targets to meet very demanding technical specifications—but the cost of a target is very high, as each one is effectively a hand-made work of art put together by highly skilled engineers and technicians. Using present-day production methods the cost and the effort involved in producing the number of targets required for a power plant would be prohibitive. Mass production techniques and factories will need to be developed to bring down the costs of capsule and hohlraum fabrication as well as methods to deliver targets accurately into the center of the target chamber. In NIF, the targets are carefully positioned and individually aligned—but in LIFE they will have to be injected like bullets, and tracked and targeted dynamically with incredibly high precision. The target chamber will be filled with xenon gas in order to capture the energetic ions released by an ignited target and to protect the first wall. This gas is expected to reach very high temperatures (typically 5000–7000°C) and the target-chamber first wall will be at about 600°C. It is claimed that studies show that a hohlraum can be designed to act as a thermal shield for the cryogenic capsule while it is in transit through this hostile environment.

Neutrons will damage the first wall and surrounding structures of the target chamber, as in magnetic confinement (Box 13.2). An important requirement for the long-term success of fusion energy will be to develop new materials—especially low-activation materials (Box 13.3)—that can withstand this damage for many years of power-plant operation. The LIFE study argues that, rather than waiting for such advanced materials to be developed, it would be preferable to build a demonstration fusion power plant as soon as possible using materials that are already available and to introduce the advanced materials progressively over the longer term. A first wall constructed from presently available steels, suitably protected by xenon gas, could withstand the required operating conditions and have a working life—determined by neutron-induced damage—of perhaps 1 year of operation at full power. The spherical geometry inherent in ICF makes chamber replacement relatively straightforward (an advantage compared to magnetic confinement).

It is argued that this approach would allow the construction and initial operation of a demonstration fusion power plant to proceed in parallel with a program of developing advanced materials that could be incorporated as they become available.

13.9 Tritium Breeding

A requirement common to both magnetic- and inertial-confinement power plants based on the DT reaction is the need to breed tritium fuel. Small quantities of tritium are available as a by-product from fission reactors (especially the Canadian CANDU reactors), and they can be used to start up a fusion power program; thereafter, each fusion plant will have to be self-sufficient in breeding its own supplies. The basic principles of the reaction between neutrons and lithium in a layer surrounding the reacting plasma are discussed in Box 4.2. This layer, known as the *tritium-breeding blanket*, is discussed in more detail in Box 13.9. The tritium will be extracted from the blanket and fed back into the power plant fuel. Lithium is widely available as a mineral in the Earth's crust. Reserves have been estimated to be equivalent to a 30,000-year supply at present world total energy consumption. Even larger quantities of lithium could be extracted from seawater.

Box 13.9 Tritium Breeding

It is necessary to breed tritium efficiently in a fusion power plant, obtaining at least one tritium atom for every one consumed (Box 4.2). There is only one neutron produced by each DT reaction, and so it is necessary to make sure that, on average, each neutron breeds at least one tritium nucleus. Inevitably, some neutrons are lost due to absorption in the structure of the power plant and the fact that it is not always possible to surround the plasma with a complete blanket of lithium—the toroidal geometry of magnetic-confinement systems restricts the amount of blanket that can be used for breeding on the inboard side of the torus.

However, each DT neutron starts with energy of 14.1 MeV, so in principle it may undergo the endothermic ^{7}Li reaction, producing a secondary neutron as well as a tritium nucleus. The secondary neutron can then go on to breed more tritium with either the ^{7}Li or the ^{6}Li reactions

$$^6\text{Li} + n \rightarrow {}^4\text{He} + T + 4.8\,\text{MeV}$$
$$^7\text{Li} + n \rightarrow {}^4\text{He} + T + n - 2.5\,\text{MeV}$$

There are other nuclear reactions, such as those with lead and beryllium, that can increase the number of neutrons.

$$^9\text{Be} + n \rightarrow {}^4\text{He} + {}^4\text{He} + 2n$$

Because of the losses by absorption, calculation of the tritium-breeding rate is complicated and needs to take into account the detailed mechanical structure and materials of the fusion power plant, the form in which the lithium breeding material is contained, the type of coolant used in the blanket, and the fraction of the blanket that can be devoted to the breeder. Account must be taken of components that reduce the space available for the breeding blanket: plasma heating, diagnostics, and divertor systems in magnetic confinement, and the capsule injection and driver systems in inertial confinement. In general, it should be easier to achieve tritium self-sufficiency with inertial confinement than with magnetic confinement because of the simpler geometry and fewer penetrations. A breeding ratio that is slightly greater than unity, say 1.01, will be sufficient in principle for self-sufficiency if the tritium is recycled promptly. Ratios between 1.05 and 1.2 have been calculated for typical blanket designs. Testing these designs and confirming the predictions are important tasks for the next-step experiments.

We have discussed the DT fusion reaction almost exclusively up to now. DT is the favorite candidate fuel for a fusion power plant because it has the largest reaction rate of all the fusion reactions and burns at the lowest temperature. The DT cycle has two principal disadvantages: (1) it produces neutrons, which require shielding and will damage and activate the structure; (2) the need to breed tritium involves the extra complexity, cost, and radial space required for the lithium-breeding blanket. There are alternative cycles for fusion fuels (see Box 13.10) that might avoid some of these problems, but these fuels are much more difficult to burn than DT and have other problems. Developing fusion power plants to burn any fuel other than DT seems to be far in the future of fusion energy.

Various chemical forms have been considered for the lithium in the blanket. Possible lithium compounds are lithium–lead or lithium–tin alloys, lithium oxide (Li_2O), lithium orthosilicate (Li_4SiO_4), and a lithium/fluorine/beryllium mixture (Li_2BeF_4) known as Flibe. Lithium and lithium–lead offer the highest breeding ratios without the use of a neutron multiplier such as beryllium. Metallic lithium can be used in principle. Its disadvantage in liquid metallic form is that it is very reactive chemically, and it also is easily set on fire in the presence of air or water in the case of an accident. There are also problems for magnetic fusion power plants in pumping conducting metals at high rates across a magnetic field.

In considering the various engineering and scientific problems of a fusion power plant, the conclusion of a great many detailed studies and independent reviews is that no insuperable problems exist. It is within our abilities to construct such a device, but it will require a sustained effort in research and development of the technology. How long this will take depends on the urgency with which we need the results—but typically this development would take about 20 years. The important questions, whether we really need fusion energy and whether it will be economical, are discussed in the next chapter.

Box 13.10 Alternative Fuels

A mixture of DT fuel burns at a lower temperature than other fusion reactions, but it has the disadvantage that it produces energetic neutrons that will damage surrounding structures and make them radioactive.

$$D + T \rightarrow n(14.1\,MeV) + {}^4He\,(3.5\,MeV)$$

There has been a great deal of interest in trying to avoid neutrons by using the fusion reaction between deuterium and the lighter isotope of helium (^{3}He), which produces a proton and ^{4}He.

$$D + {}^3He \rightarrow p(14.68\,MeV) + {}^4He\,(3.67\,MeV)$$

The D^{3}He reaction would require a temperature of around 700 million degrees and the plasma would lose most of its energy by two types of radiation (known as *bremsstrahlung* and *synchrotron radiation*) that are not generally a problem at the rather lower temperature of 200 million degrees required for DT. Also, D^{3}He would need to burn at much higher plasma density and pressure—and the required combination of temperature, density, and pressure is far beyond the expectations of any present-day magnetic-confinement system. There is a fundamental problem in obtaining a supply of fuel, because ^{3}He does not occur naturally on Earth and it is very expensive to manufacture. People have speculated enthusiastically about the idea of mining ^{3}He on the Moon, where the surface rocks have accumulated ^{3}He due to eons of bombardment by the solar wind—but in reality this is a far-fetched idea. The concentration of ^{3}He in lunar rocks is so extremely dilute that a ton of moon rock contains less energy than a ton of coal—it would require an enormous mining effort to extract any ^{3}He in worthwhile quantities and transport it back to Earth at realistic cost. The final bit of bad news is that although D^{3}He does not produce any neutrons, the deuterium in the fuel mixture would react together—and the DD reactions do produce neutrons.

As explained in Chapter 4, there are two DD reactions and they occur with equal probability. One reaction produces a proton and tritium, the other a neutron and ^{3}He.

$$D + D \rightarrow p\,(3.02\,MeV) + T\,(1.01\,MeV)$$
$$D + D \rightarrow n(2.45\,MeV) + {}^3He\,(0.82\,MeV)$$

The tritium and the ^{3}He that are produced in these DD reactions will promptly react via the DT and the D^{3}He reactions that we have discussed already—so ultimately both branches of the DD reaction produce neutrons and the problem is not avoided. It would be very attractive to use the DD fusion reactions because the fuel is readily available and inexhaustible. Unfortunately, the conditions for fusion in DD are even more demanding than for D^{3}He and would require fusion systems that are far beyond our present capability. Perhaps future generations will find ways to exploit these reactions.

There are two other fusion reactions that avoid neutron production:

$$^3He + {}^3He \rightarrow 2p + {}^4He + 12.86\,MeV$$
$$p + {}^{11}Be \rightarrow 3 \times {}^3He + 8.7\,MeV$$

However, the reaction rates for these reactions are so small that there is little prospect that they will ever yield economical fusion power.

Chapter 14

Why We Will Need Fusion Energy

Previous chapters explain the basic principles of fusion. We have seen how fusion works in the Sun and the stars, and how it is being developed as a source of energy on Earth. After many years of difficult research, terrestrial fusion energy has been shown to be scientifically feasible. The next step, as discussed in Chapter 13, is to develop the necessary technology and to build prototype fusion power plants. However, it is also important to ensure that fusion energy is not only scientifically and technically feasible, but also economical and safe and will not damage the environment—it must be of benefit to mankind. To address these concerns, we need to look beyond fusion, at the world's requirements for energy today and in the future and at the various ways in which these needs might be met.

14.1 World Energy Needs

Readers of this book who live in one of the industrially developed countries depend on an adequate supply of energy to support their way of life. They use energy to heat and cool their homes; to produce, transport, store, and cook their food; for travel to work, on holiday, and to visit family and friends; and to manufacture the vast range of goods that are taken for granted in their everyday lives. Their work and lifestyle quickly grinds to a halt whenever an electrical power outage or a shortage of gasoline interrupts the supply of energy.

During the course of the 20th century, the increasing abundance of energy has dramatically changed the way people live and work in the industrially developed countries. Energy has eliminated almost all of the labor-intensive tasks in farming, mining, and factories that our parents and grandparents toiled over—reducing the quality of their lives and in many cases damaging their health. The abundance of energy has come from burning readily available hydrocarbon fuels: coal, oil, and gas. But the situation is vastly different in the developing countries, where billions of people have barely enough energy to survive, let alone enough to increase their living standards. Today, each person in the developing world uses less than one-tenth as much energy as those who live in some of the industrialized countries (Figure 14.1) and two billion people—roughly one-third of the world's population—live without access to any electricity.

Fusion, Second Edition.

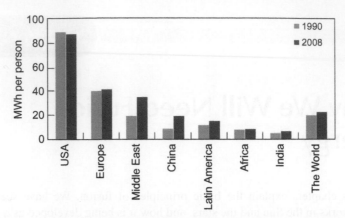

FIGURE 14.1 Average energy use by people in different parts of the world (MW hours per person per year). Each person in the US uses more than four times the world average and more than ten times as much as someone in Africa or India. Energy consumption per person in the US appears to have fallen very slightly since 1990 (although this may be due to some industries having moved from the US to other countries) but energy consumption has increased elsewhere—especially in China, where it has more than doubled over the past 20 years.

Over the past 100 years, global commercial output and the world's population have increased more rapidly than ever before, and overall annual energy consumption has risen more than tenfold. Most of the growth in commercial output and energy consumption has been in the industrialized nations, whereas most of the growth in population has been in the developing world. It took tens of thousands of years for the population of the world to reach one billion (around 1820 AD), then only about 100 years for it to double to 2 billion (around 1925), and only 50 years to double again to 4 billion (around 1975). The United Nations announced in October 2011 that the world population already had reached 7 billion and is expected to reach 8 billion by the year 2025. Assuming that there is no major catastrophe, the world population could be close to 10 billion by 2050. If people in the developing world are to achieve prosperity, their energy needs will have to be met. Finding ways of reducing our demands for energy and making more efficient use of it are important steps to reduce the overall demand—just as important as finding new sources of energy. However, even if we assume that people in the industrialized world will drastically reduce the amount of energy they use, let us say optimistically to one-half the present level, bringing the rest of the world up to this same level—and at the same time coping with the expected growth in population—will require a big increase in world energy supply.

Many studies have been made of the increase in the world population and the demand for energy, and all reach broadly similar conclusions. According to the International Energy Agency based in Paris, after the recent world economic recession, global demand for energy started to rise again in 2010 and

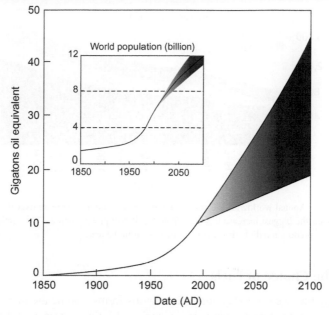

FIGURE 14.2 World primary energy consumption up to 1995, with estimated extrapolations up to 2100, for a range of scenarios. Inset is the estimate of world population growth, on which the energy growth prediction is based. The spread in the predictions is determined by the range of assumptions in the models used, from an aggressive growth policy to a very ecologically restricted one. The units of energy are 10^9 tons of oil equivalent.

is now expected to grow at a rate of 1.5% each year until 2030—which will correspond to an increase of 30% over the next two decades. The long-term projections shown in Figure 14.2 are taken from "Global Energy Perspectives to 2050 and Beyond," a report produced by the World Energy Council in collaboration with the International Institute of Applied Systems. The figures up to 1995 are actual values, and the predictions for future growth up to 2100 are based on a range of assumptions. For example, if one assumes that the global economy will continue to grow rapidly and that no constraint will be imposed on energy consumption, the energy demand in 2100 is predicted to be about four times its present level. A more conservative scenario, where both economic growth and the use of energy are reduced by strict ecological constraints, leads to an estimated doubling of demand compared to today. Such predictions inevitably have large uncertainties, but the simple message is clear—the world will use more and more energy. By the end of the 21st century, the world will be using as much energy in a period as short as 10 or 20 years as humankind has used in total from the dawn of civilization up to the present day. We have to consider what reserves and what types of fuels will be available to meet these demands for energy.

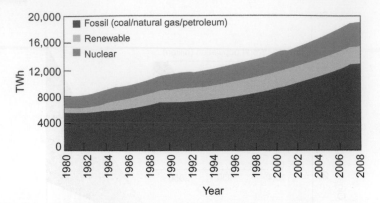

FIGURE 14.3 Annual worldwide electricity generation showing the increase over the past three decades. By far the biggest increase has been from fossil-fuel power stations. The section marked "renewable" is predominantly hydro energy, as discussed in the text.

14.2 The Choice of Fuels

The energy that we use today comes in various forms—some are more convenient than others for a particular purpose, such as heating or transport. Electricity is an important form of energy because it is clean and flexible to use and can power a wide range of sophisticated machinery, including devices like computers. Historically, the growth rate for electricity has outstripped that for other forms of energy, and the demand for electricity grows with great rapidity as a nation modernizes and its economy develops. The United States experienced a dramatic increase in demand for electricity during the first three decades of the 20th century and now a similar enhanced rate of growth is being experienced by emerging economies like those of Brazil, India, and China. About 17% of the world's energy is used in the form of electricity, but it has to be generated from a more basic form of energy. The overall average net efficiency of power plants worldwide in 2008 was about 33%—in other words, we use three units of primary energy to generate one unit of electrical energy. Worldwide (Figure 14.3), about 67% of electricity is now generated by burning fossil fuels, about 16% comes from hydro plants, and about 14% from nuclear power plants. Renewable forms of energy—geothermal, wind, solar, and combustible waste—still make only a very small, but growing, contribution: it was 0.6% in 1973 but had grown to 2.8% by 2008.

Wood and fuels like animal dung and agricultural waste were once the only significant sources of energy and remain so even today in many developing countries. These traditional fuels—nowadays known collectively as *biomass*—account for about 10% of overall world energy consumption (Figure 14.4). Many people think of these fuels as being "green" and renewable—but their widespread use has led to serious problems related to deforestation in many parts of the world. In addition, the World Health Organization estimates that 2

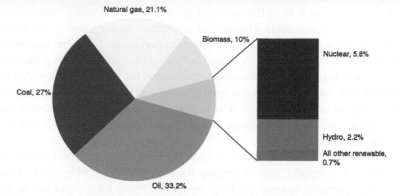

FIGURE 14.4 Global energy consumption in the year 2008. This includes the primary energy used to generate electricity. The fossil fuels (coal, oil, and gas) correspond to 81.3% of total energy consumption. The sector marked "biomass" includes wood, combustible renewables, and wastes—which of course contribute to atmospheric pollution and also to the emission of greenhouse gases, but this is offset by carbon uptake if the biomass is replanted. The expanded section shows the contribution of nuclear and hydro energy and all other renewables—including solar, wind, and geothermal energy and biofuel production.

million people die every year as a result of inhaling smoke from biomass fuels used for cooking indoors. This is twice the number of deaths caused by malaria and predominantly affects women and children in developing countries.

The exploitation of coal fueled the Industrial Revolution in Europe and America, and, together with oil and natural gas, coal provides the basis for our present-day lifestyle. The *fossil fuels*—coal, oil, and natural gas—account for about 80% of today's world energy consumption, but fossil-fuel resources are limited and will not last forever. Proven oil reserves will last between 20 and 40 years at present rates of consumption. It is widely assumed that much larger oil reserves are just waiting to be discovered—a common view is that oil companies do not find it worthwhile to explore new oil fields beyond that time horizon. But some recent studies have reached a much more pessimistic conclusion. To keep pace with the ever-increasing demand, new oil fields would have to be found in the future at a faster rate than ever before—but many of the suitable regions of the Earth have already been explored intensively. Almost all of the easily accessible oil in the world may have been discovered already (but not all the natural gas), and we may already be close to having used up one-half of the world reserves of these fuels. If this scenario proves to be true, serious oil shortages and price rises will begin in this or the next decade. There are large untapped reserves of very heavy oil in shale and so-called oil sands and there are large quantities of natural gas trapped in underground shale deposits—but it is not known if these reserves can be exploited in ways that are economically viable and ecologically acceptable. World reserves of coal are enormous by comparison and are estimated to be about 860 billion tons—sufficient to last for more than 100 years at present rates of use. One attraction of coal compared to oil and gas is that it is well distributed geographically, whereas much of the world's oil is concentrated in a handful of countries—making it very vulnerable to political factors and to price fluctuations. In particular, there are substantial amounts of coal in China and India and both countries are increasing their use of coal to help meet their rapidly growing demands for energy. Worldwide production of coal has doubled in the past decade—from about 3 billion tons per year in 1999 to around 6 billion tons in 2009—and most of this increase has been in China, which now accounts for half of the world's production and consumption of coal.

However, long before reserves of fossil fuel run out, there will be serious environmental problems due to burning them at ever increasing rates. There are strong arguments against allowing the use of fossil fuels—and in particular coal—to increase on the scale that would be needed to meet future energy demands. Burning fossil fuel produces carbon dioxide: in 2008, about 30 billion tons of carbon dioxide were released into the atmosphere—this corresponds to an average of more than 4 tons per year for every person in the world—and the amount is growing year by year. Emission of carbon dioxide doubled worldwide between 1973 and 2008. In China particularly there has been a massive increase, where emission in this 35-year period increased by 750%. Since the beginning of the Industrial Revolution, the burning of fossil fuels has contributed to an increase in the concentration of carbon dioxide in the atmosphere by 40%—from 280 ppm (parts per million) in the year 1750 to 392 ppm in 2011. Carbon dioxide (together with other gases—water vapor, methane, nitrous oxide, and ozone) causes the atmosphere to act like a greenhouse, which has a strong influence on the Earth's temperature. Without any of these gases in the atmosphere, the Earth would be on average about 33°C colder, so at optimal concentrations they do have a beneficial effect. But substantially increased amounts of greenhouse gases in the atmosphere are making the Earth hotter, which will have dramatic effects on the weather and the sea level. This is *global warming*—probably the most devastating environmental problem that human beings have created for themselves and the toughest to solve. The process and the effects are not in doubt (Figure 14.5)—all that is left to determine is by how much the Earth will heat up and how long we have before drastic action is needed. World governments have started to recognize that climate change is a serious issue: the Kyoto Protocol of 1997 and the subsequent Copenhagen Accord of 2009 have mapped out plans to reduce carbon emissions—but whether these plans go far enough and whether they will be fully implemented remain to be seen.

Burning fossil fuels causes other forms of damage to the environment, including acid rain, which is already affecting many rivers, lakes, and forests, and pollution, which damages people's health. Coal is particularly bad, because it contains many impurities (like sulfur) that cause pollution and also because it releases more greenhouse gas per unit of energy than oil or natural gas. A short-term measure undertaken in Europe during the past decade to reduce carbon emissions has been to replace coal with natural gas for

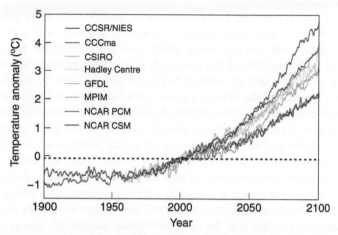

FIGURE 14.5 Global warming predictions—based on data recorded before the year 2000—showing projections of the Earth's temperature rise over the next century. There are detailed differences between the estimates of the various models, but the general upward trend is clear.

electricity generation. But this is not a long-term solution—burning natural gas also produces carbon dioxide, natural gas supplies are limited, and at best it merely puts the problem on hold for a few more years. Moreover, on a global scale, the drop in coal-fired electricity generation in Europe and elsewhere is being offset by big increases in other countries, especially China, where a new large coal-fired power plant comes into operation *every week*. Burning massive reserves of coal will be a sensible solution to future energy requirements only if new technology can be developed to burn coal without pollution and without releasing any carbon dioxide into the atmosphere. *Carbon capture* involves removing the carbon dioxide at the power plant and storing it in such a way that it does not enter the atmosphere. Some proposals involve pumping the gas into underground rock formations or deep into the oceans—but carbon capture is a new and undeveloped technology and there are many unknowns about its long-term safety and environmental effects.

Nuclear energy is clean and produces no carbon dioxide and therefore no global warming. A few decades ago there seemed to be no doubt that it would be the energy source of the future and would become the major source of electricity generation. France followed this route and now generates nearly 80% of its electricity with nuclear power plants, while Japan has reached about 30%. The US is top of the league in terms of total nuclear energy generation, although nuclear power corresponds to only about 20% of its electricity production. During the 1970s and 1980s, orders for new nuclear power plants slowed dramatically as rising construction costs (largely due to extended construction times caused by regulatory changes and litigation by pressure groups who were opposed to all things nuclear) combined with falling fossil fuel prices to make nuclear power plants less attractive economically.

The outlook for nuclear energy seemed to have become more positive in the past few years, with many countries realizing that they will need to start replacing many older nuclear power plants that are nearing the end of their working lives. It was realized that renewable energy sources could not be developed and introduced quickly enough to fill the gap that will be left when the old nuclear plants are taken out of service—and building new fossil-fuel plants would be in conflict with commitments to reduce carbon emissions. However, the accident at Japan's Fukushima Daiichi nuclear power plant in 2011 has prompted another rethink of nuclear energy policy worldwide, and Germany has already decided to close all its nuclear power plants by 2022. Nuclear energy is undoubtedly controversial and there is a longstanding debate about its use. Proponents point out that nuclear energy is a sustainable energy source that emits no greenhouse gas, that it is as safe as (perhaps even safer than) other energy sources, and that it is the only established carbon-free energy technology that can make an impact on global warming on the required time scale. Opponents argue that it poses unacceptable risks to people and to the environment. The opposition is driven as much by emotion as by hard fact, but it remains to be seen whether the relatively small risks from nuclear energy will be more acceptable than the much larger threat of global warming.

Hydroelectricity is a clean and renewable source of energy—although like all other sources of energy, it is not without risks or damage to the environment. It accounts for about 16% of electricity generation worldwide, corresponding to about 2.2% of overall energy consumption. The percentage remains fairly static in spite of some massive new hydro schemes because overall energy consumption is growing at an even faster rate. Most of the suitable rivers are already dammed in many parts of the world and there is always strong public opposition to proposals for new hydro schemes. There is no possibility that this source can be expanded sufficiently to meet future energy demands.

All other forms of renewable energy—wind, tidal, and solar—taken together contributed about 3% of the electricity generated in 2010, which corresponds to only 0.7% of the total energy supply. Although the energy in sunshine, wind, waves, and tides is enormous, there are many difficulties in harnessing these sources economically and in integrating them into a reliable supply network. At the present time, electricity generated from solar energy by photovoltaic panels is about four times more expensive than electricity generated by other methods and it has to be heavily subsidized—ultimately by the consumer. The costs of solar energy are falling as a result of improving technology and volume production—and the *SunShot* program recently announced in the United States aims to bring these costs down dramatically by the year 2020. Wind has been used as a source of energy for thousands of years to propel ships, pump water, and grind grain. The development of modern wind turbines to generate electricity started about 30 years ago in Denmark and Germany and now is a rapidly growing section of the energy industry. Worldwide, the installed capacity of wind energy increased from about 6 GW in 1996 to around

200 GW by 2010 (with Europe accounting for almost 50%) and there are projections to around 2000 GW by the year 2030. The electrical output depends very sensitively on wind speed and, since this is not constant, the average energy production is never as much as the installed capacity—typical values are about 30%. Small wind turbines can be installed in individual locations to provide energy for a specific consumer and any surplus energy can be fed into the electricity grid—but for large-scale use, large turbines are installed in *wind farms* consisting of several hundred turbines covering a large area. There is growing public opposition to new wind farms in some of the most favorable locations on land, and many new farms are being built offshore, where wind speeds are higher and the visual impact is less of a problem but costs of installation and maintenance are significantly higher.

However, there are problems with intermittent energy sources like wind and solar energy. The wind is very variable and the Sun does not shine at night. Those living in northern climes are very much aware that there are many days when the Sun does not shine at all—but we still need electricity. Electricity is very difficult and expensive to store in large quantities. At the present time the only cost-effective method of storing electricity is to run a hydro plant in reverse, pumping water from a low reservoir to one at a higher level, but there are not many suitable sites for such storage schemes. Otherwise, we have to rely on spare generating capacity that can be brought into operation quickly when the intermittent energy fails. At the present time, most countries have old fossil-fuel power plants that can provide this back-up—but when these old plants are taken out of service, new back-up plants will have to be built along with the solar and wind installations and this will add to their costs. Studies have indicated that up to 20% of total electrical energy could come from intermittent supplies and could be incorporated into a distribution grid without too much difficulty. The electrical utilities are continuing to study the effects that larger proportions of intermittent energy would have on the stability of the grid and on the cost of electricity.

Although renewable energy sources will be developed to provide a much greater proportion of the world's future energy requirements than they do at present, it seems doubtful that they will be able to satisfy the total demand. A big problem is that they are very diffuse sources of energy and deliver only about 2.5 watts per square meter of surface area. Whatever combination of renewable energy sources one takes—wind, solar, or biofuel—it has been estimated that the UK would have to use one half of its land area to generate all its energy requirements from renewable sources. There are problems of transmitting energy over large distances and big cities and centers of industry will require concentrated sources of energy. Above all, we cannot afford to rely on a single source of energy. Fusion is one of the new energy options that must be developed to give us a choice of systems that are optimally safe, environmentally friendly, and economical. The long-term solution to global warming and pollution may be to use electricity generated by environmentally friendly sources to manufacture hydrogen. Hydrogen causes no pollution when burned—it could replace fossil fuels in transport and could be stored to provide a back-up source of energy.

14.3   The Environmental Impact of Fusion Energy

One public concern about anything connected with the word "nuclear" is safety—the fear that a nuclear power plant might get out of control and explode or melt down. Fusion has a significant advantage compared to fission in this respect—a fusion power plant is intrinsically safe; it cannot explode or "run away." Unlike today's nuclear power plants, which contain relatively large quantities of uranium or plutonium fuel, enough to keep them going for many years, a fusion power plant will contain a very small amount of deuterium and tritium fuel. It is just enough to keep the reaction going for just a few seconds, and if it is not continually replaced the fusion reaction goes out and stays out.

A second safety consideration is radioactive waste. A nuclear fission power plant produces two types of radioactive waste. The most difficult to process and store are the waste products that result from the uranium fuel cycle—these are intensely radioactive substances that need to be separated from the used fuel and then stored safely for tens of thousands of years. The fusion fuel cycle produces none of these radioactive waste products—the waste product is helium gas, which is neither toxic nor radioactive. The tritium fuel itself is radioactive, but it decays relatively quickly (the half-life is 12.3 years). All of the tritium fuel that is produced will be recycled quickly and burned in the power plant. An important safety feature is that there need be no shipments of radioactive fuels or wastes into or out of the fusion plant—it would be self-contained. The raw materials necessary for the fusion fuel, lithium and water, are completely nonradioactive. There is enough lithium to last for at least tens of thousands of years and enough deuterium in the oceans to be truly inexhaustible. (Moreover, in the very long term, more advanced types of fusion may be developed to burn only deuterium.) These raw materials are widely distributed worldwide, making it impossible for any one country to monopolise the supplies.

The second source of waste from a nuclear power plant is the structure of the reactor, which is made radioactive by the neutrons emitted during the nuclear reactions. Fusion and fission are broadly similar in this respect, and some components of a fusion power plant will become intensely radioactive. Components that have to be removed for repair or replacement will be handled and stored inside concrete shields using robotic techniques. The radioactive half-life of these structural wastes is much shorter than that of the fuel wastes produced by the uranium cycle in today's nuclear power plants. At the end of its working life, a fusion power plant will be shut down and most of the equipment will be removed. The most radioactive parts will need to be protected for less than 100 years before they can be completely demolished. Moreover, their radioactivity

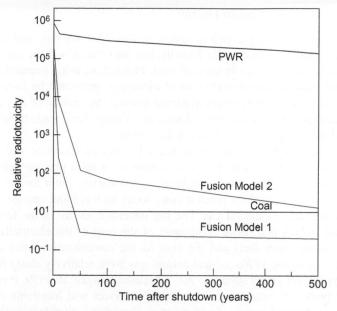

FIGURE 14.6 Comparison of the potential for harm from the radioactive waste of: a pressurized water (PWR) fission reactor, two models of fusion power plants, and a coal-fired power plant, all of the same electrical power output, plotted as a function of time after the end of operation (hazard of ingestion of radioactive products).

can be reduced by careful choice of the construction materials—research is under way to develop advanced steels and other materials that would become less radioactive, as discussed in Chapter 13. A comparison of the potential for harm from the radioactive waste of a fusion power plant after the end of its life with data from comparable fission and coal power plants is shown in Figure 14.6.

With careful design and choice of materials, the level of radioactivity left by a fusion power plant after it has been closed down for about 100 years could be comparable with that left by a coal-fired power plant. Radioactivity from a coal-fired power plant seems rather surprising. Of course, burning coal does not produce any new radioactivity, but coal contains small amounts of radioactive elements, including uranium, thorium, and potassium (as well as many other toxic elements), which are released into the environment when coal is burned. Although the concentration of uranium in coal is very small, the total quantities are surprisingly large. About 3.5 million tons of coal have to be burned every year to produce 1 GW of electricity (the requirement of a typical industrial city), and this amount of coal contains several tons of uranium. In fact, the uranium left in the coal ash is comparable to that used to supply the same amount of electricity from a nuclear power plant. Some of the uranium escapes into the air, but most is left in the ash, which is buried in landfill sites or made into blocks for building houses.

14.4   The Cost of Fusion Energy

Any new source of energy, such as fusion, has to show not only that it will be clean, safe, and environmentally friendly but also that it will be competitive with other forms of energy in terms of cost. Projections using standard energy-forecasting models show that the cost of electricity generated by fusion could be competitive when it becomes available toward the middle of this century. Several studies in the United States, Japan, and Europe have made these projections, and there is general agreement in their results.

Of course, making estimates so far ahead has big uncertainties, and they are different for each type of fuel, so comparison on a level playing field is difficult. In the case of fossil fuels, we have a good starting point for the calculations because we know how much it costs today to build and operate a power plant burning coal or natural gas. The big uncertainties are in the future cost of the fuel (which is a major component of the cost of the electricity that is produced from fossil fuel) and the cost to the environment. After substantial price rises in the 1970s, oil and natural gas were relatively cheap for more than 20 years, but prices started to rise substantially again in 1999. Predictions of future prices are bound to be uncertain, but prices will inevitably rise substantially as demand increases and reserves diminish. Other factors that have to be taken into account in making long-term forecasts include environmental concerns, which might restrict or tax the use of certain types of fuel or impose more stringent pollution standards, which would push up the costs of burning these fuels. At the present time, consumers of electricity generated from fossil fuels do not pay the cost of the resulting damage to the environment or to public health. Estimates of these external costs are difficult because the damage is widespread, is hard to quantify, and affects countries other than those that cause the pollution. Nevertheless, the external costs need to be considered in making an objective comparison between different types of fuel. Some experts calculate that the true cost of generating electricity from coal could be as much as six times higher if environmental costs were taken into account. Advanced technologies, such as carbon capture, are being developed to reduce the emission of greenhouse gas and other pollutants—but these technologies will increase the costs of building and operating the power plants.

Making predictions about the future costs of renewable sources of energy like wind, solar, and tidal energy is equally difficult. At the present time, electricity generated by these methods is generally much more expensive than that generated by fossil fuels, and, except for special locations, these methods need a government or consumer subsidy to be viable. There are also big variations from place to place. Relative costs will fall as these technologies are developed, and they are expected to become increasingly competitive as fossil-fuel costs rise. Wind and solar energy are intermittent, and the cost of energy storage or the cost of providing back-up power plants will need to be taken into account for these systems to be used as major sources of energy supply. Figure 14.7

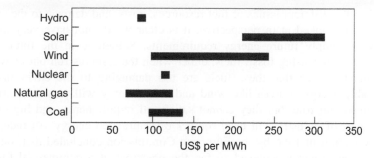

FIGURE 14.7 Estimated cost of generating electricity for new power plants entering service in the year 2016—based on a recent study by the US Department of Energy. The fossil fuels (natural gas and coal) show the range of costs with various advanced fuel cycles and with options for carbon capture; the range for wind includes estimates for onshore installations (the cheapest) and offshore (expensive); the range for solar covers photovoltaic and thermal installations and takes into account the restricted availability of the power.

gives a comparison of projected costs of generating electricity by various fuels. Hydro appears to be one of the cheapest options—but very few suitable sites remain and it can never provide anything like the capacity that is needed. The costs estimated for natural gas and coal are said to include options for carbon capture—but this new technology needs to be developed and shown to be environmentally benign and the costs need to be validated. Nuclear power is clearly very competitive on cost—but can it regain public acceptability? The renewable sources like wind and solar energy are clearly the most expensive according to present predictions—costs will no doubt fall, but there is still the problem of intermittent supply.

Predicting the costs of fusion energy involves different uncertainties. One thing about which we can be sure is that fuel costs for fusion will not be important and, moreover, will be stable. In a fusion power plant, the fuel will contribute less than 1% to the cost—a kilogram of fusion fuel will produce as much energy as 10,000 tons of coal, oil, or natural gas. The dominant part of the cost of electricity generated by fusion comes from the initial capital investment in the construction of the power plant and from the maintenance costs of replacing components during its working life. Many parts of a fusion power plant (such as buildings, turbines, and generators) will be the same as in other types of power plant, and the construction and maintenance costs are well known. The fusion plasma systems will be technically complex and therefore relatively expensive to build and maintain. Recent studies of potential designs for magnetic-confinement fusion based on the tokamak concept predict that the optimum size would generate around 1000MW of electricity at a cost that will be competitive with other fuels, and there are similar predictions for inertial-confinement fusion. Important factors in the cost of fusion electricity are the reliability of the plant and the availability with which it can be operated.

When the different issues of fuel resources, costs, and damage to the environment are looked at in perspective, it is clear we do not have very many options to supply future energy requirements. Sometime in the future we will have to stop using fossil fuels, either when the reserves run out or when governments agree that these fuels are too damaging to the environment. Renewable energy sources like wind and solar energy will play an increasingly important role, but they cannot satisfy all requirements, and big cities and centers of industry will need sources of centralized energy. An independent evaluation in 1996 by the European Commission concluded that, on the basis of the progress achieved so far, the objective of a commercial fusion power plant appears to be "a demanding but reasonably achievable goal." It will probably take at least 30 years to reach this stage. If fusion is to be ready when needed, it is important that we do not delay too long before starting to build the prototype.

Scientific Notation

The subject of fusion requires discussing very large numbers for some things, such as the density and temperature of the hot plasma, and very small numbers for other things, like the pulse length of lasers. For example, a typical plasma temperature is 100 million degrees Celsius, and a typical plasma density is 100 million million million (1 followed by 20 zeros) particles per cubic meter. At the other extreme, the pulse length of the fastest lasers is measured in terms of one-millionth of a millionth of a second or even one-thousandth of a millionth of a millionth of a second.

Very large or very small numbers are inconvenient to write in this manner, and scientists and engineers have devised a notation in which the number of zeros in large numbers is expressed as an *exponent* above the 10. Written in this way, a temperature of 100 million degrees would be 10^8 degrees, and a density of 100 million million million would be 10^{20} particles per cubic meter. A simple extension of this idea makes it easy to express very small numbers as negative exponents—thus one-tenth is written as 10^{-1}, one-thousandth as 10^{-3}, and one-millionth as 10^{-6}. Some of these subdivisions are named by prefixes attached to the unit they modify—a thousandth is indicated by the prefix *milli* and a millionth by the prefix *micro*.

This notation will already be familiar to many readers, but for others it may be a bit off-putting. In general, we have tried to spell out all the units in the main text, although this gets a bit tedious in places, but we use scientific notation in the boxes. The following table shows the equivalence between the different ways of expressing these quantities and gives examples of the ranges of time, mass, and energy that will be encountered.

In Words	Prefix	Time (s)	Mass (g)	Energy (J)	Scientific Notation
One-thousandth of a millionth of a millionth	femto-	fs			10^{-15}
One-millionth of a millionth	pico-	ps			10^{-12}
One-thousandth of a millionth	nano-	ns			10^{-9}
One-millionth	micro-	μs	μg		10^{-6}
One-thousandth	milli-	ms	mg		10^{-3}
One		s	g	J	1
One thousand	kilo-		kg	kJ	10^{3}
One million	mega-		1 Mg = 1000 kg = 1 ton	MJ	10^{6}
One thousand million (one billion in the US)	giga-			GJ	10^{9}
One million million	tera-			TJ	10^{12}
One thousand million	peta-			PJ	10^{15}

Units

Units are the measure of a particular quantity. We are all familiar with some units used in everyday life, such as degrees Celsius for temperature, cubic meters for volume, seconds for time, and amperes for electrical current. Each of these units has a precise scientific definition. The units with which we are principally concerned in nuclear fusion are temperature, density, time, and energy.

Temperature is usually measured in degrees Celsius—on this scale water freezes at 0°C and boils at 100°C. For scientific purposes, the Kelvin scale is commonly used—absolute zero on the Kelvin scale is minus 273°C. The difference between degrees Kelvin and degrees Celsius can be ignored on the scale of temperatures for fusion—these range up to hundreds of millions (10^8) of degrees Celsius. Plasma temperatures are frequently expressed in units of electron volts. For comparison, one electron volt (1 eV) is equal to 1.1605×10^4 degrees Celsius—roughly 10,000 (10^4) degrees Celsius. So temperatures in the range of millions of degrees Celsius are usually expressed in terms of thousands of electron volts (keV), and 100 million degrees Celsius is approximately 10 keV.

Density in everyday use is usually expressed in terms of mass per unit volume—we think of density as kilograms per cubic meter (or perhaps still as pounds per cubic feet). The same units could be used for plasmas—but it is more convenient to think in terms of the number of particles per unit volume rather than the mass per unit volume. Densities in a magnetically-confined plasma are typically in the range from 1 million million million (10^{18}) to 1000 million million million (10^{21}) particles per cubic meter. Inertial fusion densities are much larger and are generally expressed in terms of grams or kilograms per cubic meter.

Time is of particular importance in fusion experiments. For magnetic confinement of plasma, the confinement times have now reached about 1 second, a length of time we all feel familiar with. In inertial confinement, the times are much shorter, typically less than 1 billionth (10^{-9}) of a second.

Energy is a term that is used in a general sense—as in "fusion energy"—and as a specific scientific quantity. Energy is measured in units of joules (J). However, energies of individual particles are usually expressed in terms of the voltage through which the particle has been accelerated and 1 electron volt (eV) = 1.6022×10^{-19} joules. The total kinetic energy of a plasma is the sum of the energies of all the ions and electrons. Temperature is a measure of the average energy, and each ion and electron has average energy equal to $(3/2)kT$ (this is because a particle can move in three directions and has average energy equal to $\frac{1}{2}kT$ in each direction). Therefore, if there are n ions and n electrons per cubic meter, the total plasma energy is equal to $3nkT$.

Power is the rate at which energy is used, and it is measured in units of watts (W). One watt is equal to 1 joule per second (1 W = 1 J/s). We generally use power in discussing the output of an electrical power plant—a typical home would require a power supply on the scale of several kilowatts (kW), and a typical modern large power plant has a capacity of 1 or 2 gigawatts (GW).

ablation: The evaporation or erosion of a solid surface, usually by very-high-power heating.

Alcator C-Mod: A tokamak, at the MIT Plasma Science and Fusion Center. It is the tokamak with the highest magnetic field and highest plasma pressure in the world.

alpha particle: A helium nucleus; emitted in radioactive decay of heavy elements, and the stable nonradioactive end product of many fusion reactions.

alternative fuels: Fusion fuels other than deuterium and tritium. They have lower probability of undergoing a fusion reaction and require much higher temperature, but may produce fewer neutrons.

arcing: An electrical discharge between two surfaces in atmosphere or vacuum, driven by an electric potential. An arc can also occur between a plasma and an adjacent surface due to the potential that naturally builds up between the two. Arcing can lead to erosion of surfaces and injection of impurities into a plasma.

ARIES: A series of design studies for fusion power plants by a consortium of US laboratories, including the University of California at San Diego, the University of Wisconsin, Argonne National Laboratory, Plasma Science and Fusion Center, MIT, Princeton Plasma Physics Lab, Rensselaer Polytechnic Institute, General Atomics, and the Boeing Company.

ASDEX and ASDEX-U: Tokamaks constructed at the Max Planck Institut für Plasmaphysik (IPP) at Garching near Munich, Germany.

atom: The smallest unit of a chemical element; when it is subdivided, the parts no longer have the properties of a chemical element.

atomic mass: A, the mass of an atom relative to the standard of the carbon atom, defined as 12 units.

atomic number: Z, the number of protons in an atom.

beta: The ratio of plasma pressure to magnetic pressure—usually denoted by the symbol β.

Big Bang: The name of the theory now widely accepted as explaining the origin of the universe.

Black hole: A region of space that has so much mass concentrated in it that nothing can escape, not even light.

blanket: An annular region surrounding the plasma in a fusion power plant in which neutrons would interact with lithium to generate tritium and would be slowed down to extract their energy as heat.

Boltzmann's constant: $k = 1.38 \times 10^{-23}$ J/K (equivalent to 1.6×10^{-16} J/keV) converts a temperature in K (or keV) into energy (J).

breakeven: When the power released by the nuclear fusion reactions is equal to the external power used to heat the plasma.

breeding cycle: The reactions that enable tritium to be produced in a fusion power plant. A neutron released from the DT fusion reaction reacts with lithium in the breeding blanket around the confined plasma. This results in the production of a new tritium atom, which can be recycled and burned in the plasma. A breeding ratio of slightly greater than unity is possible.

bremsstrahlung: Electromagnetic radiation emitted by a plasma due to the deceleration of charged particles deflected by other charged particles. It can also refer to radiation due to the acceleration of a charged particle.

capsule: Small pellet of fusion fuel, usually frozen DT enclosed in a plastic shell, for inertial-confinement fusion.

carbon cycle: A chain of nuclear reactions involving carbon as a catalyst, which fuses four hydrogen nuclei into one helium nucleus, with a resulting release of energy. This cycle is important in stars hotter than the Sun.

catalyst: A substance that participates in a reaction but is left in its original state at the end.

charge exchange: A process whereby an ion and an atom exchange an electron. It is a loss mechanism when fast plasma ions are neutralized and escape from the plasma, and a plasma-forming or heating mechanism when fast atoms are injected into the plasma.

Classical Super: An early design of a hydrogen bomb, devised by Edward Teller, in which fusion reactions were supposed to be induced by the high temperature created by the explosion of an adjacent fission bomb. This design was shown to be impractical.

cold fusion: Attempts made to obtain nuclear fusion at or near room temperature. Techniques using electrolysis and sonoluminescence (bubble fusion) are generally agreed not to be substantiated. Muon-catalyzed fusion has been demonstrated scientifically but is unlikely to be able to produce net energy.

conductivity: The ability of a substance to conduct heat or electricity. Plasmas are good conductors of electricity.

confinement time: The characteristic time of energy (or particle) loss from a fusion plasma. In a plasma that is in thermal equilibrium, the energy confinement time is defined as the total energy content (MJ) divided by the total power loss (MW).

convection: The transfer of heat by motion of the hotter parts of a fluid to the colder parts.

Cosmic Microwave Background Radiation (CMBR): Microwave radiation that was created in the early life of the universe, about 380,000 years after the Big Bang.

cosmic rays: High-energy particles or nuclei that reach the Earth from outer space.

critical density: The density above which an electromagnetic wave cannot propagate through a plasma because the wave is *cut off* and reflected or absorbed.

critical mass: The mass of fissile material above which it will spontaneously undergo a nuclear fission explosion. It occurs when the material is large enough that the neutrons released can set up a chain reaction within the solid.

critical surface: The position in a plasma corresponding to the critical density.

cross-section: An effective area that describes the probability that a particular collision or reaction will occur.

Culham Laboratory: The main center for fusion research in the UK, it has a spherical tokamak, MAST. It is also the site of the Joint European Torus, JET.

current limit: The maximum current allowed in a tokamak plasma. It is determined mainly by the magnetic field and plasma dimensions.

cyclotron radiation: Radiation emitted by the plasma electrons or ions due to their cyclotron motion in the magnetic field (also called *synchrotron radiation*).

DEMO: A generic term for the next stage in the magnetic confinement program after ITER. It is still in the early conceptual design stages without any firm decisions on a construction schedule.

density limit: An empirical limit for the maximum stable density against disruptions in a magnetic-confinement system.

deuterium: An isotope of hydrogen with one proton and one neutron in the nucleus.

diagnostics: A term used in fusion research for the many different techniques of measuring the properties of a plasma, such as density, temperature, and confinement time.

DIII-D: Tokamak at the General Atomics laboratory, San Diego, CA.

direct drive: The method of heating an inertial-confinement capsule in which the energy from a laser is focused directly onto the surface of the capsule. Many laser beams have to be used to ensure uniform heating.

displacements per atom: A measure of the amount of radiation damage induced in a solid due to neutron or ion irradiation.

disruption: A gross instability of a magnetically-confined plasma that causes an abrupt temperature drop and the termination of the plasma current.

divertor: A modification of the basic toroidal magnetic-confinement system that deflects magnetic-field lines at the periphery so that particle and energy fluxes can be controlled.

DOE: Department of Energy. The US governmental department that oversees fusion energy research.

driver: The means used to provide compression and heating in an inertial-confinement experiment. Up to the present, lasers have been normally used as drivers. However, energetic ion beams have also been considered.

EAST: A tokomak built at the Institute of Plasma Physics, Hefei, China, with superconducting coils. Completed in 2006, it achieved its first plasma the same year.

electrolysis: The passing of electrical current through an electrolyte, causing ions in the electrolyte to flow to the anode or cathode, depending on their charge.

electrolyte: A solution in which electricity can flow.

electron: A particle with one unit of negative electric charge, 1/1836 the mass of a proton, which is not affected by the strong nuclear force.

electron cyclotron resonance heating: A technique for heating plasmas by applying electromagnetic radiation at the cyclotron frequency of the electrons. The frequency depends on the strength of the magnetic field and is typically 50–200 GHz.

electron volt: The energy gained by an electron when accelerated through a potential difference of 1 volt. Used as a unit of plasma temperature (1 eV corresponds to 11,600°C).

element: An atom with unique chemical properties characterized by a specified number of protons in its nucleus.

ELM: Edge-localized mode—a relaxation instability of the steep edge density gradient in the H-mode of a tokamak.

emission line: Sharp energy peak in an electromagnetic spectrum, caused by a transition from one discrete energy level to a lower one.

energy level: A quantum state of a system, such as an atom, having a well-defined energy.

EURATOM: One of the three European communities that merged into the European Union, seated in Brussels, Belgium; responsible for coordinating fusion research in Europe.

evaporation: The release of material from a surface when heated to high temperature, leading to surface erosion and injection of impurities into a plasma.

excited state: The state of an atom or molecule that has a higher energy than its ground state.

fast ignition: An approach to ICF, using an ultrashort-pulse, high-power laser to heat the plasma after compression has been achieved by other means.

FIREX: The program at the Institute for Laser Engineering at Osaka University, Japan. It is planned to develop lasers for inertial confinement by the fast-ignition method.

first wall: The solid wall directly facing the plasma in a plasma confinement experiment or a fusion power plant.

flibe: A chemical compound containing fluorine, lithium, and beryllium that has been proposed as a possible compound for tritium breeding in a fusion power plant.

fusion: In the context of this book, nuclear fusion, the process whereby light nuclei combine to form a heavier nucleus, with the release of energy.

fusion power plant: See *nuclear fusion power plant*.

GAMMA 10: A large magnetic mirror machine built in Tsukuba, Japan.

GEKKO XII: A high-power 12-beam neodymium-doped glass laser at Osaka University's Institute for Laser Engineering. Built in 1983, it is currently being upgraded for the FIREX program.

Goldston scaling: Empirical scaling for the tokamak energy confinement time.

Greenwald limit: Empirical limit for the maximum stable electron density in a tokamak.

half-life: The time required for a radioactive element to decay to half its original activity.

heavy-ion driver: A particle accelerator creating intense beams of heavy ions to compress an ICF pellet.

heavy water: Water in which ordinary hydrogen has been replaced by the heavier hydrogen isotope, deuterium.

helical winding: The twisted magnetic coils that generate the rotational transform in a stellarator.

hertz: A measure of frequency, with units of inverse seconds.

high Z: Materials and impurities like tungsten with large atomic number.

HiPER: A proposed European program to develop high-power lasers for inertial confinement. It is a consortium of 25 institutions from 25 nations.

H-mode: A regime of improved confinement in tokamaks characterized by steep density and temperature gradients at the plasma edge.

hohlraum: A cavity or can that contains an ICF capsule. The primary driver, laser, or ion beam shines onto the walls of the hohlraum, creating intense X-rays, which then compress the pellet more uniformly than the primary driver could.

hydrogen: The lightest of all elements. It consists of one proton and one electron. Hydrogen is sometimes used to refer to the three isotopes that have only one proton: hydrogen (protium), deuterium, and tritium.

hydrogen bomb, H-bomb: An atomic bomb based on fusion reactions. Deuterium and tritium undergo fusion when they are subjected to heat and pressure generated by the explosion of a fission bomb trigger.

ignition: In magnetic confinement, the condition where the alpha particle heating (20% of the fusion energy output) balances the energy losses; in inertial confinement, the point when the core of a compressed fuel capsule starts to burn.

implosion: The inward movement of an ICF pellet caused by the pressure from the rapid ablation of the outer surface by the driver.

impurities: Ions of elements other than the DT fuel in the plasma.

indirect drive: A method of heating an inertial-confinement capsule in which the energy from a laser is directed onto the inside surface of a hohlraum (or cavity) in which the capsule is contained. The hohlraum surface is heated, producing X-rays that bounce around the inside of the hohlraum, striking the capsule from all directions and smoothing out any irregularities in the original laser beams.

inertial-confinement fusion (ICF): By heating and compressing a fuel capsule very rapidly, under the right conditions, it is possible to reach sufficiently high temperatures and densities for nuclear fusion to occur before the plasma has time to expand.

instabilities: Unstable perturbations of the plasma equilibrium. Small-scale instabilities usually lead to a decrease in the plasma confinement time; large-scale instabilities can result in the abrupt termination of the plasma (see also *disruption*).

INTOR: A conceptual design study of an International Tokamak Reactor between 1978 and 1988.

ion: An atom that has more or fewer electrons than the number of protons in its nucleus and that therefore has a net electric charge. Molecular ions also occur.

ion cyclotron resonance heating (ICRH): A heating method using radio frequencies tuned to the gyro-resonance of a plasma ion species in the confining magnetic field.

ionization: The process whereby an atom loses one or more of its electrons. This can occur at high temperatures or by collision with another particle, typically an electron.

isotopes: Of an element, two or more atoms having the same number of protons but a different number of neutrons.

ITER: International Thermonuclear Experimental Reactor; an international collaboration started in 1988 to design and construct a large tokamak in order to develop the physics and technology needed for a fusion power plant. ITER is being built at Caderache in the South of France by an international consortium consisting of Europe, China, India, Japan, Russia, South Korea, and the US.

JET: Joint European Torus, currently the world's largest tokamak, built at Culham, UK, as a European collaboration. In 1997 it carried out fusion experiments with deuterium and tritium, releasing energy of 16 MW.

JT-60: Japan Tokamak; the largest Japanese tokamak, later rebuilt as JT-60U.

keV: Kilo-electron-volt; used as a unit of plasma temperature (1 keV corresponds to 11.6 million °C).

KSTAR: A tokomak built at the National Fusion Research Institute at Daejeon, South Korea, with superconducting coils. Its first plasma was achieved in 2008.

Kurchatov Institute: The Institute of Atomic Energy in Moscow, Russia's leading center for fusion research, named after Academician Igor Kurchatov. The tokamak was invented and developed here.

Larmor radius: The radius of the orbit of an electron or ion moving in a magnetic field.

laser: A device that generates or amplifies coherent electromagnetic waves. Laser light can be produced at high intensities and focused onto very small areas. Its name is an acronym for "light amplification by stimulated emission of radiation."

Laser Mégajoule Facility (LMJ): A large neodymium laser system with a planned output of 1.8 MJ being built by the French nuclear science directorate, CEA, near Bordeaux, France, and due to be completed in 2014. It is planned to use it for inertial-confinement experiments concentrating on the indirect-drive approach.

Lawrence Livermore National Laboratory: US nuclear weapons laboratory that was the leading center for mirror machines, now mainly devoted to inertial-confinement fusion being the center for the NIF program.

Lawson criterion (or condition): The condition for a fusion plasma to create more energy than required to heat it. The original version assumed for a pulsed system that the fusion energy was converted to electricity with an efficiency of 33% and was used to heat the next pulse. It is more common now, for magnetic confinement, to consider the steady-state ignition condition, where the fusion fuel ignites due to self-heating by the alpha particles.

LHD: Large Helical Device of the stellerator type built at the Japanese National Institute for Fusion Science, at Toki near Nagoya.

limiter: A device to restrict the plasma aperture and to concentrate the particle and heat fluxes onto a well-defined material surface in magnetic fusion experiments. Refractory metals, tungsten and molybdenum, as well as graphite, have been used.

linear pinch: Also known as a Z-pinch; an open-ended magnetic confinement system where the plasma current flows axially and generates an azimuthal magnetic field that constricts, or "pinches," the plasma.

L-mode: Denotes the normal confinement in tokamaks.

low-Z impurities: Material and impurities like carbon and oxygen with a relatively low atomic number.

Los Alamos National Laboratory: US nuclear weapons laboratory also engaged in fusion research.

low-activation materials: Materials used in a radiation environment that have low activation cross-sections and therefore do not get very radioactive.

magnetic axis: The field line at the center of a set of nested toroidal magnetic surfaces.

magnetic-confinement fusion (MCF): The use of magnetic fields to confine a plasma for sufficient time to reach the temperatures and densities for nuclear fusion to occur.

magnetic field lines: Directions mapping out the magnetic field in space; the degree of packing of field lines indicates the relative strength of the field.

magnetic island: A localized region of the magnetic field where the field lines form closed loops unconnected with the rest of the field lines. Magnetic islands can be formed by the application of specially designed external coils but can also form spontaneously under some conditions.

mass defect: When protons and neutrons are grouped together to form a nucleus, the mass of the nucleus is slightly smaller than the combined mass of the individual protons and neutrons.

mass spectrograph: A device for analyzing samples according to their atomic mass. The sample is ionized, accelerated in an electric field, and then passed into a magnetic field and detected at a photographic plate. The trajectory in the magnetic field is a circular path whose curvature depends on the ratio of the mass to charge of the ion. When using electrical detection, it is termed a *mass spectrometer*.

mirror machine: An open-ended system where the plasma is confined in a solenoidal field that increases in strength at the ends, the "magnetic" mirrors.

molecule: Two or more atoms held together by chemical bonds.

mu-meson, or muon: A negative particle 207 times the mass of an electron, with a lifetime of 2.2×10^{-6} s. The muon can substitute for an electron in the hydrogen atom, forming a muonic atom.

muon-catalyzed fusion: The replacement of the electron in a tritium atom by a mu meson followed by the formation of a DT molecule and a nuclear fusion reaction with the release of a neutron and an alpha particle. The mu meson is usually released after the fusion reaction, allowing it to catalyze further reactions.

NBI: Neutral Beam Injection for heating and refueling magnetic-confinement plasmas.

neoclassical theory: The classical theory of diffusion applied to toroidal systems, it takes account of the complex orbits of charged particles in toroidal magnetic fields.

neodymium glass laser: A high-power laser, used as a driver for inertial-confinement fusion, where the active lasing element is a rod or slab of glass doped with the element neodymium.

neutrino: A neutral elementary particle with very small mass but carrying energy, momentum, and spin. Neutrinos interact very weakly with matter, allowing them to travel long distances in the universe.

neutron: A neutral elementary particle with no electric charge but mass similar to that of a proton, interacting via the strong nuclear force.

neutron star: An extremely dense star composed of neutrons; believed to be formed by supernovae explosions.

NIF: National Ignition Facility, laser-driven inertial-confinement fusion experiment at Lawrence Livermore National Laboratory, completed in 2010. It has achieved an energy of 1.8 MJ, the largest laser in the world, and is being used for inertial-confinement experiments and to support nuclear weapon maintenance and design by studying the behavior of matter at high temperatures and pressures.

Nova: Was the largest laser device in the US, at the Livermore National Laboratory. Now replaced by NIF.

nuclear fission: The splitting of a heavy nucleus into two or more lighter nuclei whose combined masses are less than that of the initial nucleus. The missing mass is converted into energy, usually into the kinetic energy of the products.

nuclear force: The strong force, one of the four known fundamental forces.

nuclear fusion: The fusion of two light nuclei to form a heavier nucleus and other particles, the sum of whose masses is slightly smaller than the sum of the original nuclei. The missing mass is converted into energy, usually into the kinetic energy of the products. Because of the electric charges on the initial nuclei, a high kinetic energy is required to initiate fusion reactions.

nuclear fusion power plant: Plant where electricity will be generated from controlled thermonuclear fusion reactions. It is intended that the ITER project will be followed by a demonstration nuclear fusion power plant.

nuclear transmutation: General term used for the nuclear reactions that result in the transformation of one nucleus into another one.

nucleosynthesis: The formation of the elements, primarily in the Big Bang and in the stars.

nucleus: The core of an atom; it has a positive charge and most of the atomic mass but occupies only a small part of the volume.

ohmic heating: The process that heats a plasma by means of an electric current. Known as *joule heating* in other branches of physics.

OMEGA and OMEGA EP: Large laser facilities used for inertial-confinement experiments at the Laboratory for Laser Energetics at the University of Rochester, NY. OMEGA was completed in 1995 and uses mainly the direct-drive method for capsule compression. The OMEGA EP system will have 4 beam lines, 2 with picosecond capability.

particle-in-cell (PIC) method: A theoretical technique for studying plasma behavior by following the trajectories of all the individual particles in the fields due to the other particles. Because very large numbers of particles have to be followed, it is computationally intensive.

pellet: See *capsule*.

periodic table: A method of classifying the elements, developed by Dimitri Mendeleev in 1869. The elements are placed in rows according to atomic mass and in columns according to chemical properties.

photon: A massless uncharged particle, associated with an electromagnetic wave, that has energy and momentum and moves with the speed of light.

photosphere: The outer layer surrounding the Sun's core that emits the Sun's energy as light and heat.

pinch effect: The compression of a plasma due to the inward radial force of the azimuthal magnetic field associated with a longitudinal current in a plasma.

plasma: A hot gas in which a substantial fraction of the atoms are ionized. Most of the universe is composed of plasma, and it is often described as the "fourth state of matter."

plasma focus: A type of linear Z pinch where the plasma forms an intense hot spot near the anode, emitting copious bursts of X-rays and neutrons.

plasma frequency: The natural oscillation frequency of an electron in a plasma.

PLT: Princeton Large Torus; the first tokamak with a current larger than 1 MA.

poloidal: When a straight cylinder is bent into a torus, the azimuthal coordinate becomes the poloidal coordinate. Thus, in a toroidal pinch or tokamak, the poloidal magnetic field is the component of the magnetic field generated by the toroidal current flowing in the plasma.

positron: The antiparticle of the electron; it has the same mass as the electron but opposite charge.

Pressurized Water Reactor (PWR): A widely used type of fission reactor using pressurized water as the coolant.

Princeton: The Plasma Physics Laboratory (PPPL) at Princeton University is one of the main fusion laboratories in the US. The spherical tokamak NSTX has a small aspect ratio $R/a \approx 1.3$ and a 1 MA plasma current.

proton: Elementary particle with unit positive electric charge and mass similar to that of the neutron, interacting via the strong force.

pulsar: A source of regular pulses of radio waves, with intervals from milliseconds to seconds. Discovered in 1967 and since realized to be a rotating neutron star, a collapsed massive star with a beam of radiation sweeping across space each time the star rotates.

quantum mechanics: A mathematical theory describing matter and radiation at an atomic level. It explains how light and matter can behave both as particles and waves. One of its early successes was the explanation of the discrete wavelengths of light emitted by atoms.

q-value: See *safety factor*.

Q-value: The ratio of the fusion power to heating power in a fusion device; $Q = 1$ is referred to as breakeven.

radiation damage: When energetic particles—for example, ions, electrons, or neutrons—enter a solid, they can displace the atoms of the solid from their original position, resulting in vacancies and interstitials. Frequently, this happens in cascades. The result is usually a change in the mechanical and electrical properties of the solid. The energetic incident particles can also cause nuclear reactions, introducing gases such as hydrogen and helium as well as other transmuted atoms into the solid. The deterioration in the mechanical properties is a problem in nuclear fission power plants and needs to be taken account of in fusion power-plant design.

Rayleigh–Taylor instability: A hydrodynamic instability that can occur when two fluids of different densities are adjacent to each other, resulting in intermixing or interchange. It can be a serious problem during ICF pellet implosion.

reconnection: It is convenient to visualize a magnetic field in terms of magnetic field lines—rather like lengths of string that map out the vector directions of the field. Sometimes, magnetic field lines become tangled and sort themselves out by breaking and reconnecting in a more orderly arrangement.

red shift: The change in wavelength toward the red end of the spectrum of radiation coming from the stars and other objects in outer space. It is a Doppler shift due to the expansion of the universe.

RF heating: Radiofrequency heating; methods of heating a plasma by means of electromagnetic waves at various plasma resonance frequencies.

RFP: Reverse Field Pinch; version of the toroidal pinch where the stabilizing toroidal magnetic field reverses direction on the outside of the plasma.

Rotational transform: The angle through which a magnetic field line rotates in the poloidal direction after one complete transit around a torus in the toroidal direction.

safety factor: A measure of the helical twist of magnetic field lines in a torus, denoted by the symbol q. The value of q at the plasma edge in a tokamak must be greater than about 3 for stability against disruptions.

Sandia National Laboratories: US nuclear weapons laboratory in Albuquerque, NM, that is engaged in fusion research on both light and heavy ion drivers for inertial-confinement fusion as well as Z-pinches.

sawteeth: Periodic relaxation oscillations of the central temperature and density of a plasma in the core of a tokamak.

scaling: Empirical relations between quantities, such as the energy confinement time and various plasma parameters.

scrape-off layer: The region at the plasma edge where particles and energy flow along open field lines to the limiter or divertor.

SCYLLA: An early theta pinch at Los Alamos.

shock ignition: An ignition method in which a laser compresses the fuel capsule without heating it. The same laser is then used to generate a spherically-symmetric convergent shock wave that heats the compressed core of the capsule.

sonoluminescence: The emission of light by bubbles in a liquid excited by sound waves.

spectrograph: A device to make and record a spectrum of electromagnetic radiation.

spectrometer: A device to make and record a spectrum electronically.

spectroscopy: The measurement and analysis of spectra.

spectrum: A display of electromagnetic radiation spread out by wavelength or frequency.

spherical tokamak or torus: A toroidal configuration with an extremely small aspect ratio. The toroidal field coils of a conventional tokamak are replaced by a single large conductor through the center; the magnets are wired as half-rings off this central conductor.

Spitzer conductivity: The electrical conductivity of a fully-ionized plasma was calculated by Lyman Spitzer, who showed that it scales as $T_e{}^{3/2}$ (the temperature of the electrons to the power of 3/2). A pure hydrogen plasma with $T_e = 1.5\,\text{keV}$ conducts electricity as well as copper at room temperature.

sputtering: An atomic process in which an energetic ion or neutral atom arriving at a solid surface ejects a surface atom by momentum transfer, thus eroding the surface.

steady-state theory: A theory that accounted for the expansion of the universe by assuming that there is continuous creation of matter. It has now been replaced by the Big Bang theory.

stellarator: A generic class of toroidal-confinement systems where rotational transform of the basic toroidal field is produced by twisting the magnetic axis, by adding an external helical field or by distorting the toroidal field coils. The class includes the early figure-eight and helical winding (classical) stellarators invented in 1951 by Lyman Spitzer at Princeton as well as torsatrons, heliacs, and modular stellarators.

sticking: The process in muon-catalyzed fusion when the mu meson is carried off by the alpha particle and is therefore unable to catalyze any further fusion reactions.

stimulated emission: In lasers, the process whereby a light beam causes excited atoms to emit light coherently, thereby amplifying the beam.

superconducting: Property of a conductor to lose its resistivity at low temperature; refers to a machine with superconducting magnetic field coils.

superconductivity: The property of certain materials that have no electrical resistance when cooled to very low temperatures.

supernova: The explosion of a large star when all the nuclear fuel has been burned. Many neutrons are produced and reactions forming heavy elements occur.

T3: Tokamak at the Kurchatov Institute in Moscow, whose results in the late 1960s prompted a worldwide emphasis on tokamaks.

Teller-Ulam configuration: The design of the H-bomb in which the radiation from a fission bomb was used to compress and heat the fusion fuel. This design was used for most of the successful H-bombs.

TFTR: Tokamak Fusion Test Reactor; the largest US tokamak experiment, operated until 1997 at Princeton University. TFTR operated with deuterium and tritium, achieving a fusion power exceeding 10 MW.

thermonuclear fusion: Fusion reactions with a significant release of energy occurring in a plasma in which all the interacting particles have been heated to a uniformly high temperature.

theta pinch: A confinement system, generally open ended, where a rapidly-pulsed axial magnetic field generates a plasma current in the azimuthal (i.e., theta) direction.

Thomson scattering: Method of measuring the electron temperature and density of a plasma, based on the spectral broadening of laser light scattered from electrons.

tokamak: A toroidal-confinement system with strong stabilizing toroidal magnetic field and a poloidal field generated by a toroidal current in the plasma; proposed in 1951 by Andrei Sakharov and Igor Tamm in the Soviet Union and now the most advanced of the magnetic-confinement systems.

toroidal field: In a tokamak or stellarator, the main magnetic field component, usually created by large magnetic coils wrapped around the vessel that contains the plasma.

toroidal pinch: The first toroidal-confinement scheme, where plasma current in the toroidal direction both heats the plasma and provides the poloidal (pinch) magnetic field.

transuranic elements: Elements with mass higher than that of uranium. They are unstable and do not occur naturally, but can be made by neutron bombardment or fusing two naturally occurring elements.

triple-alpha process: A chain of fusion reactions in which three helium nuclei (alpha particles) combine to form a carbon nucleus.

tritium: An isotope of hydrogen containing one proton and two neutrons in the nucleus. It is radioactive, with a half-life of 12.35 years.

tritium breeding: See *breeding cycle*.

turbulence: In plasmas or fluids, fluctuations that can cause enhanced transport of energy and particles.

visible light: Light to which the human eye is sensitive, i.e., in the wavelength range 390–660 nanometers.

VULCAN: A versatile 8-beam 2.5 kJ laser at the Central Laser Facility of the Rutherford-Appleton Laboratory, UK. Used for inertial-confinement experiments, including the fast-ignition scheme.

wavelength: The distance over which a wave travels in one complete oscillation.

weak interaction: One of the four fundamental forces known in nature, the weak interaction is responsible for the process of beta decay, where a neutron converts into a proton releasing an electron and a neutrino.

white dwarf: The name given to the final state of a small star (mass equal to or smaller than the Sun) after it has burned all the fusion fuel.

X-rays: Electromagnetic radiation of wavelength between 0.01 and 10 nanometers.

ZETA: Zero Energy Thermonuclear Assembly; a large toroidal pinch experiment at Harwell, UK, built in 1957 and for many years the largest fusion experiment in the world.

Z-pinch: A device where a plasma is confined by the azimuthal magnetic field generated by an axial current flowing in the plasma. In a straight cylindrical pinch, z is the axial coordinate and θ is the azimuthal coordinate; in a toroidal pinch, the current is toroidal and the field poloidal.

Further Reading

There are many books on fusion energy now available, and for up-to-date information there are many websites, some of which have good introductory sections on the background physics. Here, we list some books for general background reading as well as academic textbooks and useful websites. Under the chapter headings we list books and/or websites that are particularly relevant to that chapter.

GENERAL

Bromberg JL: *Fusion: Science, Politics and the Invention of a New Energy Source*, Cambridge, MA, 1982, MIT Press. ISBN: 0262521067.

Eliezer S, Eliezer Y: *The Fourth State of Matter: An Introduction to Plasma Science*, ed 2., Philadelphia, 2001, Institute of Physics Publishing. ISBN: 0750307404.

Herman R: *Fusion: The Search for Endless Energy*, New York, 1990, Cambridge University Press. ISBN: 0521383730.

Rebhan E: *Heisser als das Sonnen Feuer*, Munich, 1992, R Piper Verlag (in German).

TEXTBOOKS

A number of textbooks have been published recently. Some of them have interesting discussions of the world energy problem and the arguments in favor of nuclear fusion, as well as detailed discussion of the physics.

Harms AA, et al.: *Principles of Fusion Energy*, 2000, World Scientific Publishing Co. Pte Ltd. ISBN: 9812380337.

Niu K: *Nuclear Fusion*, 2009, Cambridge, UK, Cambridge University Press. ISBN: 0521113547.

Freidberg JP: *Plasma Physics and Fusion Energy*, 2008, Cambridge, UK, Cambridge University Press. ISBN: 0521733170.

INTERNET SITES

There are many websites for fusion. Some of the more general and interesting ones are:
http://www.iter.org/
http://www.jet.efda.org/
http://www.efda.org/fusion/
http://fusedweb.pppl.gov/CPEP/
http://fire.pppl.gov/
http://www.nuc.berkeley.edu/fusion/fusion.html
https://lasers.llnl.gov/
http://www.lle.rochester.edu/
http://fusioned.gat.com/
http://www.iea.org/Textbase/publications/index.asp

Chapter 1—What Is Nuclear Fusion?
Wells HG: *The World Set Free*, London, 1914, Macmillan.

Chapter 3—Fusion in the Sun and Stars
Burchfield JD: *Lord Kelvin and the Age of the Earth*, Chicago, 1975, University of Chicago Press. ISBN: 0226080439 (Reprinted 1990).
Gamow G: *Creation of the Universe*, London, 1961, Macmillan.
Gribbin J, Gribbin M: *Stardust*, London, 2001, Penguin Books. ISBN: 0140283781.
Gribbon J, Rees M: *The Stuff of the Universe: Dark Matter, Mankind and Anthropic Cosmology*, London, 1995, Penguin Books. ISBN: 0140248188.
Hawking SW: *A Brief History of Time: From the Big Bang to Black Holes*, London, 1995, Bantam Press. ISBN: 0553175211.
Ronan CA: *The Natural History of the Universe from the Big Bang to the End of Time*, London, 1991, Bantam Books. ISBN: 0385253273.

Chapter 4—Man-Made Fusion
Fowler TK: *The Fusion Quest*, Baltimore, 1997, The John Hopkins University Press. ISBN: 0801854563.
Rebut P-H: *L'Energie des Etoiles: La Fusion Nucleaire Controlee*, Paris, 1999, Editions Odile Jacob. ISBN: 2-7381-0752-4 (in French).
Weise J: *La fusion Nucléaire*, Paris, 2003, Presses Universitaires de France. ISBN: 2 13 053309 4 (in French).

Chapter 5—Magnetic Confinement
Braams CM, Stott PE: *Nuclear Fusion: Half a Century of Magnetic Confinement Fusion Research*, Bristol, UK, 2002, Institute of Physics. ISBN: 0750307056.

Chapter 6—The Hydrogen Bomb
Arnold L: *Britain and the H-Bomb*, UK, 2001, Basingstoke, UK, Palgrove. ISBN: 0333947428.
Cochran TB, Norris RS: The Development of Fusion Weapons, *Encyclopaedia Britannica*: 578–580, 29, 1998.
Rhodes R: *Dark Sun: The Making of the Hydrogen Bomb*, 1995, New York, NY, Simon & Schuster. ISBN: 068480400X.

Chapter 7—Inertial-Confinement Fusion
Lindl J: *Inertial Confinement Fusion: The Quest for Ignition and Energy Gain Using Indirect Drive*, New York, 1998, Springer Verlag. ISBN: 156396662X.

Chapter 8—False Trails
Close F: *Too Hot to Handle: The Story of the Race for Cold Fusion*, London, 1990, W. H. Allen Publishing. ISBN: 1852272066.
Huizenga J.R.: *Cold Fusion: The Scientific Fiasco of the Century*, Oxford, UK, Oxford University Press. 1993, ISBN: 0198558171.

Chapter 9—Tokamaks
Wesson J: *Tokamaks*, ed 4., Oxford, UK, 2011, Clarendon Press. ISBN-10 0199592233, ISBN-13 9780199592234.

Chapter 10—From T3 to ITER

Wesson J: *The Science of JET*, Abingdon, UK, 2000, JET Joint Undertaking.

Chapter 11—ITER

http://www.iter.org/

Chapter 12—Large Inertial-Confinement Systems

Pfalzner S: *An Introduction to Inertial Confinement Fusion (Plasma Physics)*, 2006, Boca Raton, FL,Taylor & Francis. ISBN: 0750307013.

Chapter 14—Why We Will Need Fusion Energy

BP Statistical Review of World Energy, www.bp.com/statisticalreview. June 2011.

Houghton JT: *Global Warming: The Complete Briefing*, ed 2., Cambridge, UK, 1997, Cambridge University Press. ISBN: 0521629322 (ed 3, June 2004).

International Energy Agency, *Energy to 2050—Scenarios for a Sustainable Future*, 2003. ISBN: 92-64-01904-9.

International Energy Agency, "World Energy Outlook," November 2011.

MacKay DJC: *Sustainable Energy—Without the Hot Air*, 2009, Cambridge, UK, UIT Cambridge Ltd. ISBN: 978-09544529-3-3. (This is a general treatise on energy sources and the comparison between sustainable and non-sustainable energy. It makes quantitative estimates of resources and costs. It has a short section on fusion power.)

Index

The letters tb following a page number denote a text box, the letter f denotes a figure, and the letter t denotes a table.

Printed and bound by CPI Group (UK) Ltd, Croydon, CR0 4YY

03/10/2024

01040399-0005